T0233941

AI for Computer Architecture

Principles, Practice, and Prospects

Synthesis Lectures on Computer Architecture

Editor
Natalie Enright Jerger, *University of Toronto*

Editor Emerita
Margaret Martonosi, *Princeton University*

Founding Editor Emeritus
Mark D. Hill, *University of Wisconsin, Madison*

Synthesis Lectures on Computer Architecture publishes 50- to 100-page books on topics pertaining to the science and art of designing, analyzing, selecting, and interconnecting hardware components to create computers that meet functional, performance, and cost goals. The scope will largely follow the purview of premier computer architecture conferences, such as ISCA, HPCA, MICRO, and ASPLOS.

Quantum Computer System: Research for Noisy Intermediate-Scale Quantum Computers
Yongshan Ding and Frederic T. Chong
2020

A Primer on Memory Consistency and Cache Coherence, Second Edition
Vijay Nagarajan, Daniel J. Sorin, Mark D. Hill, and David Wood
2020

Innovations in the Memory System
Rajeev Balasubramonian
2019

Cache Replacement Policies
Akanksha Jain and Calvin Lin
2019

The Datacenter as a Computer: Designing Warehouse-Scale Machines, Third Edition
Luiz André Barroso, Urs Hölzle, and Parthasarathy Ranganathan
2018

Principles of Secure Processor Architecture Design
Jakub Szefer
2018

General-Purpose Graphics Processor Architectures
Tor M. Aamodt, Wilson Wai Lun Fung, and Timothy G. Rogers
2018

Compiling Algorithms for Heterogenous Systems
Steven Bell, Jing Pu, James Hegarty, and Mark Horowitz
2018

Architectural and Operating System Support for Virtual Memory
Abhishek Bhattacharjee and Daniel Lustig
2017

Deep Learning for Computer Architects
Brandon Reagen, Robert Adolf, Paul Whatmough, Gu-Yeon Wei, and David Brooks
2017

On-Chip Networks, Second Edition
Natalie Enright Jerger, Tushar Krishna, and Li-Shiuan Peh
2017

AI for Computer Architecture: Principles, Practice, and Prospects

Lizhong Chen, Drew Penney, and Daniel Jiménez

ISBN: 978-3-031-00642-5 paperback
ISBN: 978-3-031-01770-4 ebook
ISBN: 978-3-031-00067-6 hardcover

DOI 10.1007/978-3-031-01770-4

A Publication in the Springer series
SYNTHESIS LECTURES ON ADVANCES IN AUTOMOTIVE TECHNOLOGY

Lecture #55
Editor: Natalie Enright Jerger, *University of Toronto*
Editor Emerita: Margaret Martonosi, *Princeton University*
Founding Editor Emeritus: Mark D. Hill, *University of Wisconsin, Madison*
Series ISSN
Print 1935-3235   Electronic 1935-3243

AI for Computer Architecture

Principles, Practice, and Prospects

Lizhong Chen
Oregon State University

Drew Penney
Oregon State University

Daniel Jiménez
Texas A&M University

SYNTHESIS LECTURES ON COMPUTER ARCHITECTURE #55

ABSTRACT

Artificial intelligence has already enabled pivotal advances in diverse fields, yet its impact on computer architecture has only just begun. In particular, recent work has explored broader application to the design, optimization, and simulation of computer architecture. Notably, machine-learning-based strategies often surpass prior state-of-the-art analytical, heuristic, and human-expert approaches. This book reviews the application of machine learning in system-wide simulation and run-time optimization, and in many individual components such as caches/memories, branch predictors, networks-on-chip, and GPUs. The book further analyzes current practice to highlight useful design strategies and identify areas for future work, based on optimized implementation strategies, opportune extensions to existing work, and ambitious long term possibilities. Taken together, these strategies and techniques present a promising future for increasingly automated computer architecture designs.

KEYWORDS

computer architecture, artificial intelligence, machine learning, automated architecture design, design space exploration, design optimization, supervised learning, unsupervised learning, semi-supervised learning, reinforcement learning

Contents

Preface

Computer architecture has become a stage for a dramatic shift in design practices. Traditional methods that rely upon exhaustive searches and heuristic approximations are being pushed to their limits as design complexity increases. These limitations, coupled with the slowing of Moore's law, motivate a breakthrough in computer architecture design. In our view, this breakthrough comes in the form of practical artificial-intelligence-based designs.

The state-of-the-art in computer architecture has begun to reflect this promising new paradigm with a growing number of works that span practically all major architectural components. Nevertheless, existing resources for artificial intelligence (AI) and architecture design tend to focus on novel architectures to support AI models, essentially architecture for AI rather than AI for architecture. In our writing process, we originally sought to fulfill this need with a literature review, which included a brief background and analysis. Along the way, we decided that this growing paradigm warranted a more detailed resource that could serve as an introduction for a broader audience, particularly those that are eager to begin experimenting with AI in their own work. This book expands the original literature review to include significantly more background material, detailed case studies, and additional insights throughout the text.

Chapter 1 sets the stage for AI in architecture with a brief perspective on the growing need for alternative design strategies and the opportunities offered by AI-based designs. We continue to develop intuition on these opportunities in Chapter 2 while connecting fundamental AI principles to brief architectural examples, thus building a foundation for later chapters. These principles are quickly put into practice in Chapters 3 and 4 as we explore the vast range of AI applications in architecture and then examine three case studies based on prevalent AI-based design approaches. These case studies are intended to provide deeper insight into recent work using supervised learning, reinforcement learning, and unsupervised learning to address challenging architectural problems. With all these applications in mind, Chapter 5 offers a more critical perspective on practical considerations that may guide future work. This analysis involves high-level choices such as model selection as well as some task-specific optimizations for data collection and training overhead. The book culminates in Chapter 6 as we highlight promising opportunities for future work.

We hope that AI-based design continues to flourish and that this book encourages new practitioners to embrace increasingly automated architecture design.

Lizhong Chen, Drew Penney, and Daniel Jiménez
October 2020

Acknowledgments

The authors thank Michael Morgan and Natalie Enright Jerger for their support throughout the entire process of preparing this book as well as Jieming Yin and Joshua San Miguel for their valuable feedback.

Lizhong thanks Timothy Pinkston for his encouragement on developing a book on this emerging topic. Lizhong also greatly appreciates his family members for their unconditional love and support. Without their understanding and sacrifice, this work would not be possible.

Drew thanks Dr. Lizhong Chen for his wisdom as a professor and his inspirational role as an academic advisor. Drew also thanks his father, Bruce, for sharing his surprising AI and architecture insight despite being an analog guy, and his mother, Shirley, and his brother, Sean, for being an oasis of sanity when life is too crazy.

Daniel thanks Calvin Lin, his former dissertation advisor at the University of Texas, for believing in his ideas on machine learning in microarchitectural prediction, and Rajendra Boppana, who taught his graduate computer architecture class in 1993, for introducing him to the fascinating problem of branch prediction.

Lizhong Chen, Drew Penney, and Daniel Jiménez
October 2020

CHAPTER 1

Introduction

1.1 THE RISE OF AI IN ARCHITECTURE

In the past decade, artificial intelligence (AI) has rapidly become a revolutionary factor in many fields, with examples ranging from commercial applications, such as self-driving cars, to medical applications, including improved disease screening and diagnosis. AI application in architecture has, likewise, undergone a transition from being a theoretical novelty to being a driving force behind design, control, and simulation in practically all components. This incredible proliferation that we now observe ultimately finds its foundation somewhat earlier, which is where our story of AI in architecture begins.

Many early works on AI trace their origin back to the perceptron [1], proposed by Frank Rosenblatt in 1958. This perceptron model, as we will discuss further in Chapter 2, represented a simple, yet crucial, step toward developing a machine that possesses human-like learning capabilities. Naturally, there were some limitations in this early work: most notably, this perceptron could only solve linearly divisible functions, thus it was not possible to solve XOR or XNOR functions [2]. Nevertheless, research continued on. In the following decades, more sophisticated models that could learn any function emerged as promising successors. Multi-layer models, in particular, rose to the forefront along with the backpropagation algorithm. The growing complexity of these models demanded more efficient algorithms, larger and more diverse datasets, and, of course, greater processing capabilities. This history culminates with the recent surge in AI research and applications in diverse fields, along with historic breakthroughs in long-standing tasks such as image classification (as shown in Table 1.1). Another notable example is AlphaGo [3], which learned to play the game Go. Humans have played for the game for thousands of years, yet AlphaGo surpassed the world champion after just weeks of training.

Advancements in computer architecture design complexity and processing capabilities have, to some extent, mirrored those in AI. The first practical integrated circuits were developed in the late 1950s and commercialized in the 1960s [11]. A decade later, Intel introduced the first microprocessor, Intel 4004, using just 2300 transistors and addressing 640 byte of memory. Consistent growth predicted by Moore's law[1] produced increasingly complex processors that, by the 1990s, comprised millions of transistors. In 2001, IBM introduced Power4, the first dual-core processor on a single die, roughly marking the start of the current many-core era. Shortly after, Intel released the Pentium 4 featuring simultaneous multi-threading, an industry first for general-purpose processors. Architecture design history culminates in recent years with

---

[1]Moore's law states that the number of transistors incorporated on a chip will double approximately every two years [12].

Table 1.1: Image classification progression (ILSVRC2012 dataset [4])

Model	Error (top-5)	Layers	Parameters
LeNet-5 (1998) [5]	x	7	60 K
AlexNet (2012) [6]	15.3%	8	60 M
VGGNet (2014) [7]	7.3%	19	138 M
GoogLeNet (2014) [8]	6.7%	22	4 M
ResNet (2015) [9]	3.6%	152	60.3 M
Noisy Student (2019) [10]	1.3%*	x	480 M
* This number is not directly comparable to others due to differences in evaluation procedure, but is included for rough comparison as the current state-of-the-art.			

the slowing of Moore's law and the end of Dennard scaling;[2] chips now contain billions of transistors, yet only a fraction of these can be active due to thermal constraints. These limitations have placed increasing burden on architects to supplant technology scaling with architectural advances. Consequently, there is a growing need for more sophisticated strategies for design, control, and optimization, thus setting the stage for AI application in architecture.

Traditionally, the relationship between computer architecture and AI has been relatively imbalanced, focusing on architectural optimizations to accelerate AI algorithms. In fact, the recent resurgence in AI research is, in part, attributed to improved processing capabilities. These improvements are enhanced by hardware optimizations exploiting available parallelism, data reuse, sparsity, etc. in existing AI algorithms. Nevertheless, within the past decade, this paradigm has begun to shift toward a more symbiotic relationship as more and more works successfully apply AI to architecture (see Figure 1.1).

Notable early works include perceptron-based branch prediction [14] and reinforcement-learning-based self-optimizing memory controllers [15], which demonstrated the potential for AI-based methods to surpass more traditional schemes, even in well-studied components. The incredible proliferation following these successes has resulted from a variety of factors:

- the fundamental applicability of AI (discussed in Section 1.3) enables diverse applications, ranging from performance prediction to hardware-based malware detection [16]; applications in emerging technologies have been particularly successful at addressing both standard objectives (e.g., power and performance) and non-standard (e.g., lifetime and tail-latency) objectives;

- continued advances in AI algorithms and learning capabilities encourage novel design strategies and applications in previously infeasible tasks; design space exploration, a

---

[2]Dennard scaling refers to constant transistor power density due to reduced voltage/current with reduced transistor size [13].

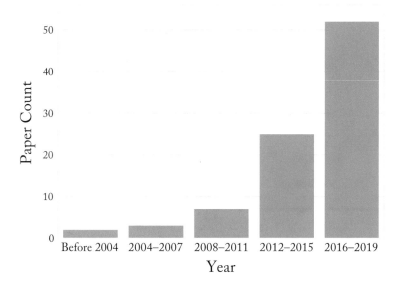

Figure 1.1: Publications on AI applied to architecture (for works examined in Chapter 3).

task that has traditionally relied upon heuristics, is now possible in immense spaces surpassing 10^{100} [17]; and

- reduced overhead of AI models via task-specific optimizations has produced increasingly sophisticated online applications involving real-time performance and power prediction; for example, novel preemptive control strategies have been shown to improve reliability (e.g., dynamic error correction in networks-on-chip (NoCs) [18]), reduce energy, provide better utilization of existing resources (e.g., dynamic resource sharing between latency-critical and best-effort workloads), and much more.

These nascent works, although limited, establish the broad applicability and auspicious future for AI-enabled architectural design. Existing AI-based approaches, ranging from dynamic voltage and frequency scaling (DVFS) with simple classification trees to design space exploration via deep reinforcement learning (DRL), have already surpassed their respective state-of-the-art human expert and heuristic-based designs. AI-based design will likely continue to provide breakthroughs as promising applications are explored.

1.2   THE SCOPE OF AI

The study of AI began with the notion of constructing machines that can think and act using knowledge and understanding in the same way that we, as humans, do in our daily lives. These

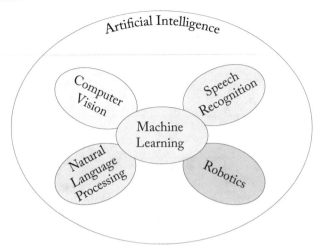

Figure 1.2: AI and example subfields.

early philosophical concepts have, over time, burgeoned into a broad range of subfields, with a select few represented in Figure 1.2.

Speech recognition explores methodologies to convert from spoken to written language. Natural language processing similarly involves building an understanding of language, enabling capabilities such as automatic translation (e.g., Google Translate). AI applications in robotics seek to enhance control and decision-making by learning from experience rather than painstakingly crafted routines. Computer vision, a related subfield, enables machines to interact and gain information from the real world based on visual data. Finally, at the center is machine learning, which has been rapidly adopted not just in other AI subfields, but also in architecture. In fact, machine learning application to architecture will be our primary focus in later chapters. Nevertheless, we do not intend for the reader to assume that AI in architecture will always focus on such strategies. As an example, future work may perhaps apply natural language processing and automated code generation to architecture design, allowing a design/simulation tool to carry out automated verification based on written goals.

1.3   FUNDAMENTAL APPLICABILITY

Machine learning has been rapidly adopted in many fields as an alternative approach for a diverse range of problems. This fundamental applicability stems from the powerful relationship-learning capabilities of machine learning algorithms. Specifically, machine learning models leverage a generic framework in which models learn from examples, rather than explicit programming, enabling applications in many tasks, including those too difficult to represent using standard programming methods.

Using this generic framework, there may be many possible approaches for a given problem. For example, in the case of predicting instructions per cycle (IPC) for a processor, one can experiment with a simple linear regression model, which learns a linear relationship between features, such as core frequency and cache size, and the prediction target, IPC. This approach may work well or it may work poorly. In the latter situation, one can try different features, non-linear feature combinations or a different model entirely, with another common choice being an artificial neural network (ANN). The diverse options offered by various models, model parameters, and training features allow machine-learning-based approaches to match the task at hand.

Even with this generally applicable framework, we are still left with the following questions.

1. Why are these capabilities applicable to architectural designs?

2. Why are traditional methods not sufficient?

We answer the first question by considering the general process of architectural design. Drawbacks or shortcomings in existing designs lead to ideas, which we can investigate and put into practice. Along the way, we may explore a wide variety of implementation strategies to optimize performance, power, area, etc. This procedure naturally involves significant time simulating designs and optimizing implementation parameters, both of which become increasingly challenging in complex designs. Architecture design tasks such as these are ideal for machine learning applications. Specifically, with machine learning, important metrics such as performance and power can be approximated to a high degree of accuracy using data from less than 1% of the design space. Design optimization can also be simplified by machine learning models that automatically explore design trade-offs and thereby learn underlying contributors to performance, power, etc. These capabilities and more (covered in Chapter 3) have already shown substantial promise throughout architecture design.

The second question, regarding sufficiency, can be answered by examining differences between traditional methods and machine learning methods. In this discussion, and for the remainder of this book, we use *traditional methods* to broadly encompass heuristics as well as schemes with more dynamic behavior (e.g., control-theoretic methods) but comparatively limited "learning" capabilities.

Diversity in machine learning approaches, as discussed in the previous IPC prediction example, starkly contrasts with heuristics, which have long been a primary tool in architectural design. In general, heuristic-based approaches rely upon a single strategy working well under common circumstances and working at least adequately under uncommon circumstances. This approximation is, in turn, necessitated by substantial complexity in individual sub-systems and in interactions between sub-systems. There are, however, no guarantees that these approximations will hold for a new application, let alone an alternative system configuration. In fact, numerous works in Chapter 3 highlight the shortcomings in heuristics due to their inherently static nature, thus limiting adaptation to new behaviors. Machine learning approaches differ in that they do not require assumptions about the data they will observe, yet can still guarantee convergence (i.e., determine a beneficial approximation, policy, etc.) in dynamic environments.

Generalizability is another significant differentiating factor between traditional and machine learning methods. For example, consider computational predictors, such as the stride prefetcher. These predictors can "learn" in the sense that prefetching decisions are updated at runtime to better accommodate changing workload behavior. Regardless, these predictions are ultimately limited by assumptions regarding particular memory access patterns; specifically, stride prefetchers assume repeating deltas between accessed values, resulting in comparatively poor prediction accuracy on access patterns without this behavior. Context-based predictors (e.g., state-of-the-art TAgged GEometric (TAGE) branch predictors [19] and neural branch predictors [20]) instead make predictions by correlating recorded outcomes (stored in history tables) and the current branch, thereby enabling accurate predictions on a broader range of workloads. Early TAGE predictors [21] were indexed using just the program counter and global branch history, while neural branch predictors have been shown to efficiently select useful information from a variety of features, offering a multi-perspective view [20]. In theory, a machine-learning-based prefetcher can use these diverse features to dynamically optimize itself for any function (i.e., memory access pattern). More recent TAGE predictors [19] incorporate a "statistical corrector," essentially a neural predictor indexed using a variety of features, to achieve state-of-the-art accuracy. Notably, a meaningful portion of branches that are mispredicted by TAGE predictors are correctly predicted by alternative neural predictors,[3] indicating benefits from generalized approaches that can learn from many features.

Similar arguments for improved generalizability can also be made when comparing machine learning and control-theoretic approaches. Generally speaking, control-theoretic approaches rely upon ground-truth models relating the operating environment to more holistic values that can be used to identify deviations from a target state. When a high-accuracy ground-truth model cannot be constructed, control-theoretic approaches can no longer provide formal guarantees and may fail to deliver satisfactory behavior [22]. In fact, a number of prior works [22, 23] emphasize difficulties in formally modeling all of the complex, dynamic interactions that occur between various architectural resources. Machine learning techniques provide a natural solution to this problem by implicitly learning the "ground-truth" model through sampled data while offering inherent resistance to noisy measurements. Perhaps more significantly, machine learning approaches can enable more effective control by predicting appropriate behaviors rather than simply reacting. We note, however, that the formal guarantees offered by some traditional methods remain attractive for a number of applications. For that reason, hybrid schemes integrating both traditional methods and machine learning capabilities have grown in popularity. Regardless, the significant improvements observed in many recent works largely

---

[3] Assuming practical overhead for both predictors.

hinge on ease of implementation and improved generalizability to diverse applications through machine learning methods.

Limitations in traditional methods are also observed in design space exploration, thus representing another compelling application for machine learning. In practice, exhaustive exploration becomes intractable in sufficiently large spaces, notably in architectural design where there may be hundreds of useful parameters, each with potentially hundreds of configurations. Conventional techniques may attempt to prune away subspaces with sub-optimal configurations and optimize around high-performing configurations. Given sufficient time, these conventional approaches can generally locate appropriate configurations. Nevertheless, these conventional approaches may be too slow for practical application. Machine-learning-enabled design space exploration directly addresses this concern by exploiting patterns or relationships between configurations to predict beneficial subspaces, ultimately resulting in a more efficient and intelligent exploration procedure. A notable example is the AlphaGo model [3], which efficiently explores a design space of approximately 10^{170} [24]. As discussed by Young [25], many problems in computer architecture can be optimized via reward functions; although models such as AlphaGo were not originally intended for architecture, their principles can be directly applied in diverse applications.

Replacing traditional methods in control, prediction, and design space exploration tasks is, ultimately, just a subset of the possibilities for machine learning applications in architecture. As we will see in Chapter 3, applications span diverse components, goals, and implementation requirements, yet are all enabled by one common element: machine learning.

1.4 LEVELS OF AI FOR ARCHITECTURE

Diverse AI approaches have already produced state-of-the-art advancements for many tasks and practically all major architectural components. Nevertheless, the current scope of AI in architecture is still nowhere near its full potential; computer architects must still determine bottlenecks or areas for improvement, develop implementation strategies that take advantage of AI, and define specific goals for design or optimization. Naturally, these current limitations also represent future opportunities: continuous advances in AI algorithms, coupled with more efficient hardware implementation, may facilitate increasingly automated designs. We categorize these applications into four levels to foster understanding of both current work and future possibilities.

Level 0: No Automation
Level 0 represents the conventional architectural design strategy based on human knowledge and traditional approaches. Practically all work fell within this category until approximately 2001, when AI-based branch prediction was introduced. Even now, this is still the predominant design approach.

Level 1: AI-assisted Design

At level 1, AI replaces some traditional design tools, particularly heuristics and analytical models. Application involves specific design tasks, such as design space exploration or performance prediction, generally in isolation (i.e., no knowledge of the architecture as a whole). These models still rely upon human guidance to define the problem and any relevant information. We consider this to be the highest level of AI application in current work.

Level 2: AI-guided Design

At level 2, AI becomes the primary mechanism to perform more generalized design tasks involving many architectural components. Human intervention is required primarily to set optimization goals (e.g., for specific use cases or operating environments). These models may draw from decades of human design examples to explore novel architectural configurations while applying prior capabilities.

Level 3: AI-dominated Design

This last level can, perhaps, be regarded as the end goal, representing a paradigm in which architectural innovations are explored, implemented, and optimized by machine learning algorithm(s) with little to no human intervention. AI models become capable of understanding the function of architectural components, thus enabling radical architectural innovation such as transistor-level design space exploration. There are, of course, numerous challenges to address before this paradigm can be made reality. Nevertheless, we consider this promising frontier to motivate broader implementation and innovative AI applications.

CHAPTER 2

Basics of Machine Learning in Architecture

Our exploration of AI for architecture begins by considering the fundamental machine learning principles upon which most works are founded, whether in design, control, simulation, or other tasks. Throughout this chapter, we emphasize an architecture-focused perspective with the intent to build intuition about practical machine learning applications. We therefore intersperse necessary mathematical insights with high-level architectural examples, all while focusing on learning approaches and models with proven applicability. Nevertheless, for the sake of completeness, we introduce additional approaches that may eventually find practical application. Low-level implementation details will be considered later in Chapter 5.

In general, there are four main categories of learning approaches to consider: supervised learning, unsupervised learning, semi-supervised learning, and reinforcement learning. These approaches can be differentiated by *what* data is used and *how* that data is used to facilitate learning. Similarly, many appropriate models may exist for a given problem, thus enabling significant diversity in applications based on the learning approach, hardware resources, available data, etc.

2.1 SUPERVISED LEARNING

In supervised learning, training involves input features and output targets, with the result being a model that can predict the output for new, previously unseen inputs. These models can handle output that is both continuous/numerical, referred to as regression, or discrete/categorical, referred to as classification. Common regressions tasks include performance or power prediction while common classification tasks include resource configuration and workload scheduling.

2.1.1 MODELS

Supervised learning models can be generalized into four categories: decision trees, Bayesian networks, support vector machines (SVMs), and artificial neural networks (ANNs) [26].

Decision Trees
Decision trees are best understood by their structures, which organize data into a tree. In a decision tree, each internal node represents a feature and each branch represents possible values

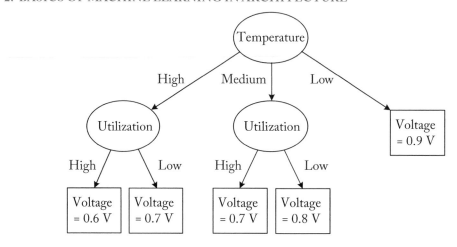

Figure 2.1: Example decision tree for dynamic voltage control.

of the feature represented by the parent node. Leaves represent the result, which could be either a value or a label depending upon the supervised learning task. At runtime, new inputs are evaluated by traversing the tree based on the input feature values.

Consider the example decision tree illustrated in Figure 2.1. Observe that there are two features: temperature and utilization. Leaves provide labels for voltage values ranging from 0.6–0.9 V. We could imagine that this decision tree may be used in an on-chip core DVFS task, where voltages are lowered to reduce peak core temperature. Evaluation begins at the root. If current temperature readings are high, traversal follows the leftmost branch. From there, if current resource utilization is low, traversal follows the right branch and arrives at the leaf "Voltage = 0.7 V." This result could then be sent to a controller for actuation.

Decision tree complexity is largely specified by tree depth and the desired precision for branches. Specifically, deeper trees can include more features, thus providing greater capabilities to differentiate otherwise identical inputs. Similarly, more branches per node can enable finer granularity and help differentiate particular feature values. The aptly named *random forest* model combines a number of standard decision trees, each trained with different features, thus providing a third dimension to adjust model complexity.

Bayesian Networks

Bayesian networks, like decision trees, follow a well-defined graphical structure involving nodes and edges. These nodes and edges, however, have significantly different meanings in the two models. Bayesian networks embed conditional relationships into their graphical structure; nodes represent random variables and edges represent conditional dependence between these variables. Although random variables may incorporate the same recognizable features as with decision

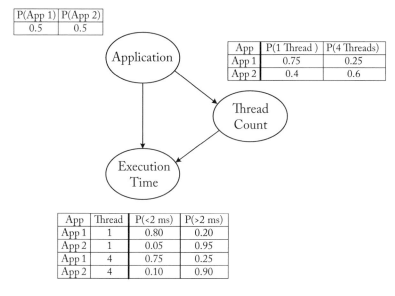

P(App 1)	P(App 2)
0.5	0.5

App	P(1 Thread)	P(4 Threads)
App 1	0.75	0.25
App 2	0.4	0.6

App	Thread	P(<2 ms)	P(>2 ms)
App 1	1	0.80	0.20
App 2	1	0.05	0.95
App 1	4	0.75	0.25
App 2	4	0.10	0.90

Figure 2.2: Example Bayesian network for execution time prediction.

trees, we interpret their possible values as probability distributions. Evaluation therefore considers the distribution of random variables that impact the variable of interest.

Figure 2.2 presents an example Bayesian network with three random variables: application, thread count, and execution time. Execution time can take a broad range of values depending upon both the application and the thread count. In other words, execution time is conditioned upon the application and thread count. Thread count is similarly conditioned upon the application as some applications cannot be parallelized or exhibit diminishing returns with increasing parallelism. Bayesian networks make use of these conditional relationships to represent the probability of specific outcomes/value for one variable using the others (as seen in the tables alongside each node). For example, the probability of an application using four threads given that we are running application 1 is $P(4\ \textit{Threads} \mid \textit{App}\ 1) = 0.25$.

Let E represent execution time, T represent thread count, and A represent the application. Using this notation, we can succinctly represent the joint distribution over all variables as

$$P(E, T, A) = P(E|T, A)\,P(T|A)\,P(A). \tag{2.1}$$

If we have information about some variables (e.g., thread count and execution time), we can infer the probability for others (e.g., application) via Bayes rule [27]:

$$P(A|E,T) = \frac{P(E,T,A)}{P(E,T)}$$

$$= \frac{P(E,T,A)}{\sum_{A'} P(E,T,A')} \tag{2.2}$$

$$= \frac{P(E|T,A)\,P(T|A)\,P(A)}{\sum_{A'} P(E|T,A')\,P(T|A')\,P(A')}.$$

Say we want to predict if the last executed application was application 1. Further, assume that we know the execution time and thread count was $E \leq 2\,ms$ and $T = 1$. Plugging in these values, we find that

$$P(A = 1|E \leq 2\,ms, T = 1) = \frac{P(E \leq 2\,ms|T = 1, A = 1)\,P(T = 1|A = 1)\,P(A = 1)}{\sum_{A'} P(E \leq 2\,ms|T = 1, A')\,P(T = 1|A')\,P(A')}$$

$$= \frac{0.8 * 0.75 * 0.5}{0.8 * 0.75 * 0.5 + 0.05 * 0.4 * 0.5} \tag{2.3}$$

$$= 0.968.$$

Intuitively, we can make sense of this result since execution time is, in this example, usually less than 2 ms when we are running application 1, but rarely less than 2 ms for application 2, especially when thread count is low.

In this simple example, the network structure (i.e., the relationships between the three variables) is known. Likewise, all three variables are observable, meaning that we can run simulations and record the values for all variables. In practice, it may not be realistic to restrict ourselves to a known structure and/or observable variables. In fact, the reader can likely imagine scenarios in which it may be difficult (if not impossible) to explicitly measure and represent all underlying factors affecting execution time. In such circumstances, Bayesian networks can expand to include *latent* or *hidden* variables.

Figure 2.3 illustrates a modified Bayesian network in which the observable application variable has been replaced with a latent variable. This latent variable is, in contrast to the observable variables, learned during training, thus allowing the network to discover conditional relationships rather than relying solely upon prior knowledge. Recent work [28] adopts a similar approach, using observations from diverse applications to learn a distribution for latent variables affecting execution time. The learned relationships are then applied to make predictions for a new application.

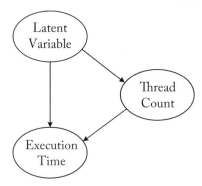

Figure 2.3: Example Bayesian network with latent variable.

Support Vector Machines (SVMs)

SVMs are generally known for their function rather than for a particular graphical structure. Specifically, SVMs identify an optimal hyperplane[1] that maximizes the distance, also known as the *margin*, between examples from all classes/groups and the hyperplane. This division is specified by the training examples that fall closest to the hyperplane. These examples are referred to as the *support vectors*. Predictions for new data points are then made based on their position with respect to the learned support vectors.

Figure 2.4 illustrates these concepts for two separate classification approaches on a two-class dataset. The first approach, seen in Figure 2.4a, represents a possible result using a simple classification model that updates the dividing line only when an example is classified incorrectly. Although this approach successfully classifies all training examples, the margin between classes is relatively small.

At first glance, it may not be clear why we care about the margin if all training examples can be classified correctly. To illustrate the significance of margin, consider Figure 2.4b, which includes a new data point (shown with a dashed outline) that we would like to classify. For the sake of discussion, assume that this data point actually belongs to the green class (i.e., is most similar to examples on the right side of the figure, so is shaded green). Clearly, this new data point falls on the left side of the hyperplane so, by this simple classification model, would be misclassified. In contrast, Figure 2.4c highlights the results from SVM classification with an optimized dividing line. Now, comparing against the added test point in Figure 2.4d, we see that the new dashed point is correctly classified as it falls on the right side of the hyperplane. In practice, the optimized margin should provide greater classification accuracy for new data. This difference between the simple classification approach (Figures 2.4a,b) and SVM classification (Figures 2.4c,d) stems from the problem formulation. In the following, we present a general

---

[1]The term *hyperplane* refers to a general, possibly high-dimension, division in space. For example, hyperplanes in 2D space are lines (1D).

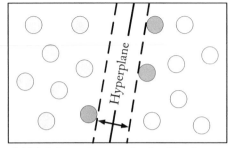

(a) Training using simple classification approach. The margin (shown by the arrows) is relatively small.

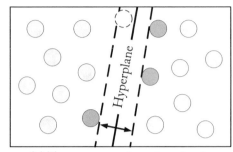

(b) Test on new point using simple classification approach.

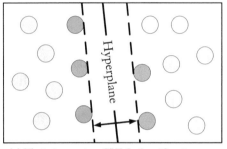

(c) Training using SVM classification. The margin (shown by the arrows) is maximized.

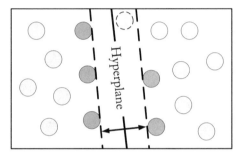

(d) Test on new point using SVM classification.

Figure 2.4: Comparison of classification models.

and simplified mathematical derivation using vector notation (i.e., $\vec{x}$) to accommodate data and weights with arbitrary dimensions.

Any k-dimensional hyperplane can be written in the form

$$\vec{w} \cdot \vec{x} - b = 0, \tag{2.4}$$

where $\vec{w}$ is the normal vector and b is a bias term that allows the hyperplane to be offset from the origin. Specifically, this offset is given as

$$\frac{b}{\|\vec{w}\|} \tag{2.5}$$

or, in words, b divided by the magnitude or L2-norm of $\vec{w}$. The points $\vec{x}$ that satisfy this equation are, therefore, the points that fall on the hyperplane.

Now, consider our supervised learning setup. We are provided with a set of n training examples $\mathcal{X} = \{(\vec{x}_1, y_1), \ldots, (\vec{x}_n, y_n)\}$, where each $\vec{x}_i$ is a k-dimensional data point and y_i is the training label (-1 or 1), denoting the class to which the data point belongs.[2] Given these examples, we would like a method that reliably maximizes prediction accuracy for new examples.

Looking back at Figure 2.4c, we can define two additional hyperplanes (shown as the dashed lines) that are parallel to each other and the original hyperplane given by Equation (2.4). Assume that data points on the right side have a label of $y_i = 1$ and data points on the left side have a label of $y_i = -1$. These new hyperplanes are

$$\vec{w} \cdot \vec{x} - b = 1 \tag{2.6}$$

and

$$\vec{w} \cdot \vec{x} - b = -1 \tag{2.7}$$

representing the right and left hyperplanes, respectively. The distance between these two hyperplanes is then given as

$$\frac{2}{\|\vec{w}\|}. \tag{2.8}$$

It follows that the distance between these hyperplanes can be maximized by minimizing $\|\vec{w}\|$. Correct classification using this setup requires that all points with label $y_i = 1$ fall to the right side of its hyperplane, or

$$\vec{w} \cdot \vec{x}_i - b \geq 1. \tag{2.9}$$

Similarly, all points with label $y_i = -1$ should fall to the left side of its hyperplane, or

$$\vec{w} \cdot \vec{x}_i - b \leq 1. \tag{2.10}$$

We can then combine Equations (2.9) and (2.10), along with the constraint on $\vec{w}$ to arrive at the overall SVM problem formulation, given as

$$\begin{aligned} \text{minimize} \quad & \frac{1}{\|\vec{w}\|} \\ \text{subject to} \quad & y_i(\vec{w} \cdot \vec{x}_i - b) \geq 1, \ i = \{1, \dots, n\}. \end{aligned} \tag{2.11}$$

The $\vec{w}$ and b that satisfy these constraints can then be used to classify new examples: for an input x, the predicted label will be $sgn(\vec{w} \cdot \vec{x} - b)$, producing either $+1$ or -1.

This binary classification setup with an SVM can be practically applied to, perhaps, surprisingly complex tasks. For example, one work [29] applied this model to predict the relative performance when executing a code segment on a CPU or a GPU. Training data in this application could include characteristics such as the dynamic instruction mix or relevant hardware performance counters while training labels could be either 1 or -1, denoting whether execution time is lower or higher on a GPU compared to a CPU. Given this setup, it should make sense

---

[2]This derivation uses a setup with two classes for simplicity. SVM classification can also be extended to multi-class problems. One formulation builds k binary classifiers for k classes, each trained to determine if the new example is from its class (vs. the others).

that maximizing the margin between classes (i.e., being very confident about whether GPU execution will be faster) could be beneficial; converting code to execute on a GPU takes time, thus we want to be sure that this work will pay off.

Note that the previously derived binary classification setup assumes that the dataset is indeed linearly separable and that the optimization problem has a solution, which may not always be the case. A modified formulation involving an error term ξ (called a *slack* variable) is given as

$$
\begin{aligned}
\text{minimize} \quad & \frac{1}{\|\vec{w}\|} + C \sum_{i=1}^{n} \xi_i \\
\text{subject to} \quad & y_i(\vec{w} \cdot \vec{x}_i - b) \geq 1 - \xi_i, \ \xi_i \geq 0, \ i = \{1, \ldots, n\}.
\end{aligned}
\tag{2.12}
$$

The parameter C introduces a trade-off between the tolerated error and the margin of the model.

The formulation for regression using support vector machines is similar, but introduces an additional slack variable, with the optimization problem as follows:

$$
\begin{aligned}
\text{minimize} \quad & \frac{1}{\|\vec{w}\|} + C \sum_{i=1}^{n} (\xi_i + \xi_i^*) \\
\text{subject to} \quad & y_i - (\vec{w} \cdot \vec{x}_i - b) \leq \xi_i, \\
& (\vec{w} \cdot \vec{x}_i - b) - y_i \leq \xi_i^*.
\end{aligned}
\tag{2.13}
$$

These two separate slack variables allow the upper and lower bounds of the regression function to be bounded independently. In this formulation for regression, C provides a trade-off between the tolerated error and the model complexity. An important result of further derivation (shown without proof for simplicity) provides an equivalent form, given as

$$
y = \sum_{i=1}^{n} (a_i - a_i^*)(x_i \cdot x_i) + b,
\tag{2.14}
$$

where a and a^* are Lagrange multipliers. This result demonstrates that, as mentioned earlier, regression (as well as classification) can be performed using just the support vectors x_i determined during training.[3] SVMs can also be extended to nonlinear problems using kernel methods [31].

Artificial Neural Networks
Artificial neural networks (ANNs), or simply neural networks, represent a broad category of models that are defined by their structures, which are reminiscent of neurons in the human brain; layers of nodes are connected via links with learned weights, enabling particular nodes to respond to specific input features.

An example perceptron model, illustrated in Figure 2.5, represents the simplest ANN, comprising just one layer with a single node. Here, we show a perceptron with three inputs

---

[3]See Smola and Schölkopf [30] for further discussion.

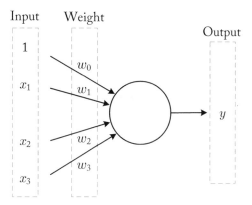

Figure 2.5: Perceptron model.

$\mathcal{X} = \{x_1, x_2, x_3\}$, their three corresponding link weights $\mathcal{W} = \{w_1, w_2, w_3\}$, and a bias term with weight w_0. In the general case with n inputs, the output y is calculated as the weighted sum of the inputs, given by Equation (2.15):

$$y = w_0 + \sum_{i=1}^{n} w_i x_i. \tag{2.15}$$

Finally, predictions can be made based on whether y is above or below a threshold value. The bias term is therefore included to shift the weighted sum relative to the threshold value.

More complex deep neural networks (DNNs), such as the model in Figure 2.6, include many layers of these weighted sums. Note that we show three nodes per layer purely for illustrative purposes. In reality, the numbers of nodes per layer is highly variable: the number of nodes in the input layer is determined by the number of input features; the number of nodes in the output layer is decided by the task characteristics, such as the number of possible classes in a classification task; finally, the number of nodes in the hidden layer can be any value. Wider hidden layers increase both overhead and network capacity (i.e., the ability of the network to approximate complex functions). The threshold function of the perceptron is additionally generalized here as a nonlinear activation function. Weighted sums at each node are passed through a nonlinear activation function, then sent as input to the next layer. These multi-layer models, when combined with nonlinear activation functions, can approximate effectively any continuous function (according to the universal approximation theorem [32]).

As an example, consider the scenario presented for decision trees involving on-chip DVFS. In that example, the decision tree used temperature and utilization as features. In addition, we may want to include the amount of data traffic, which could provide further insight into utilization trends. The model will, therefore, have three inputs, represented by $\mathcal{X} = \{x_1, x_2, x_3\}$. At the output, each node represents one possible voltage (e.g., 0.6 V, 0.7 V, and 0.8 V). After

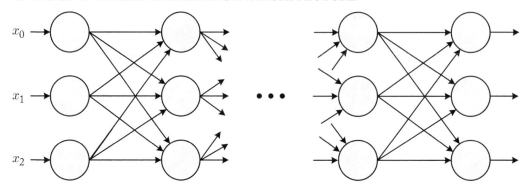

Figure 2.6: Artificial neural network.

several layers of weighted summations and nonlinear activations, results are generated at the output layer as a final weighted sum and nonlinear activation. The node in the output layer with the highest activation value corresponds to the predicted label that, in the same manner as with the decision tree prediction, can be sent to a controller for actuation.

Additional neural network variants such as convolutional neural networks (CNNs) and recurrent neural networks (RNNs) introduce alternative structures that are better suited for specific data types. CNNs are particularly useful for data processing involving spatial locality, such as images. In these models, data is processed using convolution operations, in which *convolutional filters* are applied to regions of the input (e.g., 3×3 groups of pixels). Each convolutional filter is trained to recognize a particular arrangement of input values, such as an edge or a curve in an image. Straightforward application to architecture is possible by representing data, such as communication between cores, as matrices or heatmaps, thus allowing patterns to be identified and optimized [33]. More generally, these capabilities can be applied to tasks such as prefetching, where convolutional filters can be trained to detect specific patterns of instructions and branch history. We will also see a more detailed example later in the case studies. RNNs are instead specialized for temporal data by integrating loops from output to input, thereby encouraging sequence recognition capabilities. Ongoing research has, however, shown that simply adding a loop from output to input can be problematic. Specifically, information may be lost during forward propagation and updates may be unstable during backpropagation [34]. One solution is the long short-term memory (LSTM) model. At a high level, the main difference between a standard RNN and an LSTM model is that the LSTM maintains an internal *cell state*, which can store information separately from the output, and uses gates to control information flow between the input, output, and internal state. These improved LSTM models have been successfully applied to a variety of prefetching tasks, most notably for their resistance to noisy data patterns and for long-term pattern recognition.

General Considerations

All these supervised learning models can be used in both classification and regression tasks, although there are some distinct high-level differences. Variants of SVMs and neural networks tend to perform better for high-dimension and continuous features and also when features may be nonlinear [26]. These models, however, tend to require more data compared to Bayesian networks and decision trees.

There are additional considerations that apply when online training is required. Many works presented in Chapter 3 consider online updates solely for neural networks and decision trees since individual updates are relatively inexpensive. Online updates are possible for other models, but are much less common. LASVM [35], for example, provides an algorithm to perform online SVM training. Updates using LASVM require several searches with complexity based on the number of support vectors, whereas updates for neural network have fixed complexity relative to the network size. Ultimately, these guidelines are just a starting point and, as such, alternative models can always be considered for a specific application.

2.1.2   THE LEARNING PROCESS

In general, neural networks have become the most commonly applied machine learning model in architectural tasks, especially in tasks necessitating online training. Although diverse neural network variants (e.g., DNNs, CNNs, and RNNs) exhibit some variations in training due to their structure, all variants can be understood through the synergistic processes of forward and backpropagation.

Forward Propagation

Formally, forward propagation is the process through which input to the network is converted to an output, usually involving a series of weighted summations and activation functions. This may sound familiar since we have, in fact, already observed a simplified version of forward propagation with the DVFS example from the prior section. The truth is that forward propagation is required for both inference (i.e., generating a prediction/classification) and training; during inference we seek to generate an output to apply to the task at hand while during training we seek to adjust the neural network to produce more accurate results.

Consider the forward propagation procedure in Algorithm 2.1, with key values illustrated in Figure 2.7. As expected, calculation primarily involves multiplication between weights and inputs from the previous layer. The activation function following each layer can be considered arbitrary here since its primary purpose is enabling nonlinear function approximation, which is possible with any nonlinear activation function.[4] The predicted output from the network (line 6) is the last step required for inference. Line 7, involving both the network output and the target output, is included during training. In the calculation on line 7, there are two pieces that we have not yet discussed: the loss function and the regularizer. The loss function generates a value

---

[4]In practice, the nonlinear activation function can impact training behavior and model accuracy.

Algorithm 2.1 Forward propagation and cost function computation for a multi-layer neural network.

Input : Network depth $\mathcal{D}$
Input : Input data features $\mathbf{X} = \{x_1, \ldots, x_n\}$
Input : Weight vectors $\mathbf{W}^{(i)}$, $i \in \{1, \ldots, \mathcal{D}\}$
Input : Bias vectors $\mathbf{B}^{(i)}$, $i \in \{1, \ldots, \mathcal{D}\}$
Input : Target output value(s) $\mathbf{Y} = \{y_1, \ldots, y_m\}$
Output : Total cost $\mathbf{J}$

Let $\mathbf{A}^{(i)}$, $i \in \{1, \ldots, \mathcal{D}\}$ be the input to the activation function at layer i
Let $\mathbf{H}^{(i)}$, $i \in \{0, \ldots, \mathcal{D}\}$ be the output of layer i
Let $\mathcal{F}$ represent an arbitrary activation function
Let $\widehat{\mathbf{Y}}$ represent an the network output value(s)
Let $\mathcal{L}$ represent an arbitrary loss function
Let $\mathcal{R}$ represent an arbitrary regularizer
Let Θ represent the network weights/biases

1: $\mathbf{H}^0 = \mathbf{X}$
2: **for** $j \in \{1, \ldots, \mathcal{D}\}$ **do**
3:     $\mathbf{A}^{(j)} = \mathbf{W}^{(j)}\mathbf{H}^{(j-1)} + \mathbf{B}^{(j)}$
4:     $\mathbf{H}^{(j)} = \mathcal{F}(\mathbf{A}^{(j)})$
5: **end for**
6: $\widehat{\mathbf{Y}} = \mathbf{H}^{(D)}$
7: $\mathbf{J} = \mathcal{L}(\mathbf{Y}, \widehat{\mathbf{Y}}) + \lambda\mathcal{R}(\Theta)$

---

based on the difference between the generated output and the target output. An exemplar is the mean squared error, given as

$$MSE = \frac{1}{N}\sum_{i=1}^{N}(y_i - \hat{y}_i)^2. \tag{2.16}$$

The regularizer provides an optional method to control the learning capabilities of the network by enforcing an additional penalty. In practice, we usually penalize just the link weights, not the biases. We may, for instance, want to push weights toward zero to eliminate some calculations. This simplification can be achieved, as shown in Equation (2.17), by taking the L1 norm (i.e., summing over the magnitude of all weights) such that a network with more non-zero terms has a higher cost, thus creating a tendency toward zero-valued terms. This approach is referred to as either L1-regularization or lasso regularization. Additional options to simplify

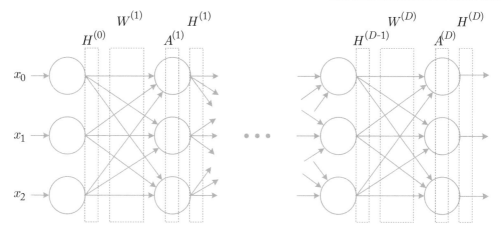

Figure 2.7: Decomposition of artificial neural network values.

the network are discussed in Section 2.1.3:

$$\lambda \mathcal{R}(\Theta) = \lambda ||W|| = \lambda \sum_{j=1}^{m} |w_j|. \tag{2.17}$$

Taken together, the loss function and the regularizer represent the overall *cost function*, which is the input during backpropagation.

Backpropagation

Following forward propagation, we are left with a cost function, which indicates how well our model fits the training data and any other criteria that we impose (e.g., the regularizer). Additionally, recall that we did not actually alter the network during forward propagation; if we again provided the same input, we would observe the same output. The actual "learning" is done during backpropagation.

At its core, backpropagation in neural networks acts to adjust link weights based on their impact on the cost. Intuitively, weights that contributed significantly to the cost should be corrected most, while weights that contributed little to the cost do not require significant modification. We can achieve this correction using the gradient of the cost function with respect to any weight (or bias) in the network, written as

$$\nabla_{\mathbf{W}_i}\mathbf{J}. \tag{2.18}$$

This calculation is, however, complicated by the fully-connected and multi-layer nature of many neural networks. As a result, the partial derivative with respect to a particular weight layer must consider the weighted summations and activation functions in all of the subsequent layers. This structure can be represented as a combination of partial derivatives, based on the calculus chain rule, to walk backward through the network. Re-using notation from Algorithm 2.1, we have

$$\nabla_{\mathbf{W}_i}\mathbf{J} = \frac{\partial \mathbf{J}}{\partial \widehat{\mathbf{Y}}} \frac{\partial \widehat{\mathbf{Y}}}{\partial \mathbf{A}^{(D)}} \frac{\partial \mathbf{A}^{(D)}}{\partial \mathbf{H}^{(D-1)}} \cdots \frac{\partial \mathbf{H}^{(i)}}{\partial \mathbf{A}^{(i)}} \frac{\partial \mathbf{A}^{(i)}}{\partial \mathbf{W}^{(i)}}. \tag{2.19}$$

Algorithm 2.2 Backpropagation and updating weight/bias values in a multi-layer neural network.

Input : Network depth $\mathcal{D}$
Input : Weight vectors $\mathbf{W}^{(i)}$, $i \in \{1, \ldots, \mathcal{D}\}$
Input : Bias vectors $\mathbf{B}^{(i)}$, $i \in \{1, \ldots, \mathcal{D}\}$
Input : Target output value(s) $\mathbf{Y} = \{y_1, \ldots, y_m\}$
Input : Total cost $\mathbf{J}$

Let $\mathbf{G}$ be the error gradient vector
Let $\mathbf{A}^{(i)}$, $i \in \{1, \ldots, \mathcal{D}\}$ be the input to the activation function at layer i
Let $\mathbf{H}^{(i)}$, $i \in \{0, \ldots, \mathcal{D}\}$ be the output of layer i
Let $\mathcal{F}$ represent an arbitrary activation function
Let $\widehat{\mathbf{Y}}$ represent an the network output value(s)
Let $\mathcal{L}$ represent an arbitrary loss function
Let $\mathcal{R}$ represent an arbitrary regularizer
Let Θ represent the network weights
Let $\circ$ denote the Hadamard (elementwise) product

1: $\mathbf{G} = \nabla_{\widehat{\mathbf{Y}}}\mathbf{J} = \nabla_{\widehat{\mathbf{Y}}}\mathcal{L}(\mathbf{Y}, \widehat{\mathbf{Y}})$
2: **for** $j \in \{\mathcal{D}, \ldots, 1\}$ **do**
3:     $\mathbf{G} = \nabla_{\mathbf{A}^{(j)}}\mathbf{J} = \mathbf{G} \circ \mathcal{F}'(\mathbf{A}^{(j)})$
4:     $\nabla_{\mathbf{W}^{(j)}}\mathbf{J} = \mathbf{G}\mathbf{H}^{(j-1)\top} + \lambda\nabla_{\mathbf{W}^{(j)}}\mathcal{R}(\Theta)$
5:     $\nabla_{\mathbf{B}^{(j)}}\mathbf{J} = \mathbf{G} + \lambda\nabla_{\mathbf{B}^{(j)}}\mathcal{R}(\Theta)$
6:     $\mathbf{G} = \nabla_{\mathbf{H}^{(j-1)}}\mathbf{J} = \mathbf{W}^{(j)\top}\mathbf{G}$
7: **end for**

We detail the backpropagation process in Algorithm 2.2, again using matrix notation to succinctly represent an entire layer at once. The error is propagated backward at each layer (line 3) by taking the Hadamard (elementwise) product with the derivative of the activation function. Intuitively, the derivative of the activation function tells us how quickly the output $\mathcal{F}(\mathbf{A}^{(i)}) = \mathbf{H}^{(i)}$ was changing with respect to the pre-activation values $\mathbf{A}^{(i)}$. Therefore, using the error gradient from layer i, the derivative of the activation function tells us how quickly the error was changing with respect to the pre-activation values $\mathbf{A}^{(i)}$. Lines 4 and 5 perform the weight/bias updates for the current layer. Focusing on line 4, the update for weights involves the vector product of the current error with the output from the prior layer. This may seem counterintuitive, yet recall we want the partial derivative with respect to the weights. These weights were originally calculated as

$$\mathbf{A}^{(j)} = \mathbf{W}^{(j)}\mathbf{H}^{(j-1)} + \mathbf{B}^{(j)} \tag{2.20}$$

so taking the gradient with respect to the weights, we are left with

$$\nabla_{\mathbf{W}^{(j)}}\mathbf{A}^{(j)} = \mathbf{H}^{(j-1)}. \tag{2.21}$$

A similar result holds for the bias update (line 5). Both updates are further adjusted, as necessary, to account for the regularizer. It naturally follows that, in order to continue propagating the error gradient backward through the network, we must take the gradient with respect to the output from the prior layer (line 6), giving

$$\nabla_{\mathbf{H}^{(j-1)}}\mathbf{A}^{(j)} = \mathbf{W}^{(j)}. \tag{2.22}$$

In general, we observe that backpropagation, like forward propagation, primarily requires matrix multiplication. Many works exploit this fact by re-using the same hardware for both forward and backpropagation, thus enabling practical online updating.

2.1.3 FEATURE SELECTION

Supervised learning, as well as semi-supervised learning (discussed in Section 2.3), involves additional consideration for feature selection. As we saw in the decision tree DVFS example, features such as temperature and utilization should be chosen carefully since the model must learn to predict solely based on values for those feature. Consequently, approaches for feature selection can substantially impact model performance, including concerns such as overfitting and computational overhead, as well as more abstract concerns, such as feature interpretability. In some works, feature selection is entirely based on expert knowledge. Additional, more general, approaches can either supplant or supplement expert knowledge. Here, we briefly describe the three general approaches. Example use cases will be examined in Chapter 4 and further compared in Chapter 5.

One set of approaches, called *filter methods*, considers features individually using metrics involving statistical correlation or information theoretic criteria such as mutual information. These approaches are usually the least computationally intensive so may be preferred for very large feature sets, but model performance may be sub-optimal since evaluation criteria in filter methods do not consider feature context [36]; two features that provide little benefit individually may be beneficial together. Many alternative approaches therefore consider feature subsets.

Wrapper methods provide a black-box approach for feature selection by directly assessing the performance of a learning model [36]. Commonly applied greedy approaches include forward selection and backward elimination. In forward selection, features are progressively added to selected feature subset based on improvement to the overall learning model. Conversely, backward elimination progressively removes features that provide little benefit. Genetic searches, in the context of feature selection, likewise fall within this category.

Embedded methods integrate feature selection into the learning model to provide a trade-off between filter and wrapper methods [37]. Regularization is a widely used embedded method that allows the learning model to be fit, while simultaneously forcing feature coefficients to be small (or zero). Features with zero coefficient values can then be removed. This method obviates the need for iterative feature selection present in wrapper methods, which can have high computational requirements [36].

2.2   UNSUPERVISED LEARNING

The supervised learning methods covered in the prior section rely upon labels or target values that serve as a reference for correct model behavior. Naturally, there may be tasks for which it is difficult to determine an appropriate label, potentially due to task complexity or required human effort. In these situations, we may instead consider unsupervised learning techniques that require just input data to extract information without human effort. As we will see, these techniques can be applied to reduce dataset dimensionality or discover patterns in a dataset that may not be obvious to humans. We begin by discussing a few common techniques, then move on to an example.

2.2.1   TECHNIQUES

Thus, far, the primary two unsupervised learning techniques applied to architecture are principal components analysis (PCA) and k-means clustering. These two techniques and their variants are also quite commonly used in other fields, likely due to their relative simplicity in implementation.

Principal Components Analysis

The primary objective of PCA is to generate an alternative representation for a dataset based on linear feature combinations [38]. In other words, PCA performs *feature extraction* by generating new features from the original data as opposed to feature selection methods discussed in Section 2.1.3, which use existing features. Naturally, both feature extraction and feature selection can be used to reduce dataset dimensionality, thereby reducing computational complexity and helping to mitigate overfitting (assuming the new features are applied to a supervised learning model). PCA-based feature extraction, however, offers a more robust approach for dataset visualization. In fact, PCA often tends to be the preferred method since it provides a straightforward method to determine two or three directions with the greatest distinction between data points. These capabilities have been applied to numerous architecture research papers that explore similarities in workload behaviors, as in Zheng et al. [39]. Application to supervised learning problems is somewhat less straightforward; PCA, in contrast to feature selection methods, does not consider the relationship between features and a target. Consequently, the criteria to remove further information and reduce dataset dimensionality may not be optimal and the resulting model performance may suffer [40]. This difference will become apparent as we examine PCA and the principles behind it.

We begin with a derivation based on Shalizi [41] to provide an intuitive motivation for PCA and its procedure. Given a dataset with dimensionality p, the goal is to determine an alternative representation with dimensionality q such that $q < p$ while retaining as much information as possible. In other words, we seek to project the dataset onto q directions in a way that minimizes the average (mean-square) distance between the original vectors and their projections. Now, assume a scenario where we are projecting onto a one-dimensional unit vector $\vec{w}$. The mean squared error when projecting an individual vector $\vec{x}_i$ can be written as

$$||\vec{x}_i - (\vec{w} \cdot \vec{x}_i)\vec{w}||^2. \tag{2.23}$$

Assuming a zero mean for vector $\vec{x}$ and knowing that $||\vec{w}||^2 = 1$ (for a unit vector), Equation (2.23) can be simplified since

$$\begin{aligned} ||\vec{x}_i - (\vec{w} \cdot \vec{x}_i)\vec{w}||^2 &= (\vec{x}_i - (\vec{w} \cdot \vec{x}_i)\vec{w}) \cdot (\vec{x}_i - (\vec{w} \cdot \vec{x}_i)\vec{w}) \\ &= ||\vec{x}_i||^2 - 2(\vec{w} \cdot \vec{x}_i)^2 + (\vec{w} \cdot \vec{x}_i)^2||\vec{w}||^2 \\ &= ||\vec{x}_i||^2 - (\vec{w} \cdot \vec{x}_i)^2 \end{aligned} \tag{2.24}$$

so then the mean squared error (MSE) across all projections can be expressed as

$$MSE(\vec{w}) = \frac{1}{n} \sum_{i=1}^{n} (||\vec{x}_i||^2 - (\vec{w} \cdot \vec{x}_i)^2). \tag{2.25}$$

Notice that only the second term involves $\vec{w}$, thus the MSE can be minimized by maximizing

$$\frac{1}{n} \sum_{i=1}^{n} (\vec{w} \cdot \vec{x}_i)^2. \tag{2.26}$$

Finally, the MSE can be broken into two components, representing the square of the mean and the variance, given as

$$\frac{1}{n} \sum_{i=1}^{n} (\vec{w} \cdot \vec{x}_i)^2 = (\frac{1}{n} \sum_{i=1}^{n} \vec{w} \cdot \vec{x}_i)^2 + Var(\vec{w} \cdot \vec{x}_i). \tag{2.27}$$

The first term (the square of the mean) will be zero following assumptions of a zero mean for $\vec{x}_i$. The overall goal of minimizing MSE for the projections is therefore equivalent to maximizing the variance of the projections along the direction $\vec{w}$.

In a general q dimensional space, the directions of maximal variance are given as the eigenvectors of the covariance matrix for the dataset [42]. These calculations are the core of PCA and can be broken into the steps as follows.

Step 1: Preprocessing

As noted above, the MSE is minimized by projecting onto the directions of maximal variance only when all variables/features in the dataset are zero-mean. There is, however, some choice when deciding whether to fully standardize the data (zero-mean *and* unit variance).[5] In general, it is recommended to fully standardize the data when features do not use identical units for measurements and especially when measurements for distinct features vary significantly [42]. We proceed with this case in the following steps given that architectural applications (e.g., analysis using hardware counters) tend to exhibit this behavior.

Step 2: Covariance matrix

Next, the covariance matrix is calculated for the transformed dataset from Step 1. Assuming a dataset matrix $\mathbf{Z}$ with n features and m readings for each feature, the covariance matrix $\mathbf{S}$, calculated as

$$\mathbf{S} = \mathbf{Z}^T \mathbf{Z}, \tag{2.28}$$

will be a square matrix with dimensions n by n.

Step 3: Eigendecomposition

It can be shown [41] that the eigenvectors of the covariance matrix $\mathbf{S}$ will point in the directions of maximal variance. When performing PCA, these directions are referred to as the principal components. Detailed calculations are omitted here as eigendecomposition is a standard calculation supported by many libraries (e.g., Scipy [43] in Python).

Step 4: Principal component selection

The previous step also produces a corresponding eigenvalue for each eigenvector. Here, the magnitude of these eigenvalues represents the degree of variation of the data along that direction. After sorting n principal components according to their eigenvalues, we can select a subset of k principal components based on the proportion of variance explained by those components as follows:

$$\frac{\lambda_1 + \lambda_2 + \cdots + \lambda_k}{\lambda_1 + \lambda_2 + \cdots + \lambda_n}. \tag{2.29}$$

Step 5: Projection

In the final step, the original dataset is projected onto each of the selected principal components, thus producing a simplified representation.

---

[5]It may seem unintuitive to standardize data and then calculate directions of maximal variance. Note, however, that standardization occurs only in the directions of the original axes, not necessarily the directions of maximal variance determined in later steps.

There are several important factors to consider when using PCA for dimensionality reduction. First, it is worth restating that the features produced by PCA are combinations of the original features so are not guaranteed to retain any physical meaning. As such, it becomes much more difficult to interpret model behaviors or learn from predictions. Additionally, as shown in Equation (2.29), substantial dimensionality reduction is at odds with information loss, potentially limiting model performance in supervised applications since information is dropped without considering its relation to a supervised target. Another drawback is that the transformed dataset may still require all of the original features to be collected, which is problematic for online architectural applications with tight overhead constraints. Regardless, we note that alternative methods [44] have been proposed to resolve some of these issues, although these methods have not (to our knowledge) been adopted in existing architectural work.

K-Means Clustering

K-means clustering, in contrast to PCA, generates an alternative dataset representation based on *discrete* labels, rather than projections onto continuous principal components. As we will discuss, these discrete labels are generated through an iterative process that groups data points with similar feature values. This procedure, although seemingly quite different from PCA, can be intuitively understood to solve a very similar problem. Specifically, both methods attempt to minimize the MSE between the original dataset and the alternative representation or, equivalently, the error when reconstructing the original dataset from the alternative representation [45]. The difference is the k-means enforces a categorical constraint on the representation for data points, with each point belonging to exactly one cluster, rather than allowing combinations of features as with PCA.

The discrete labels produced by k-means clustering (as well as other clustering variants) have proven useful in a variety of architectural design tasks. As mentioned earlier, these labels can be re-interpreted as target values for supervised learning models, thereby enabling supervised learning in otherwise unsuitable applications. In particular, performance prediction involving new workloads can use k-means to determine similarities in scaling behavior with previous examples [46]. Alternatively, some works have directly applied k-means to solve inherently cluster-related tasks in architecture. A detailed example of this scenario is presented in Section 4.3.

T-SNE

T-distributed Stochastic Neighbor Embedding (t-SNE) [47] is a comparatively newer approach for dataset visualization that, while similar in purpose to PCA, can differ in practice. Specifically, t-SNE is a *nonlinear* technique that is designed to reduce dataset dimensionality while maintaining local clustering behavior for nearby points in both high and low dimensions. This behavior is achieved by converting point-to-point distances into probabilities that are highest for close points and lower for distant points and then minimizing the Kullback–Leibler divergence, which is a measure for the similarity of two probability distributions. The resulting behavior can

diverge substantially from typical linear methods, such as PCA, which focus on preserving sep-
aration between more distant data points [47]. For that reason, t-SNE can be particularly useful
in locating structure within a group of similar data points. Furthermore, t-SNE helps avoid
crowding (i.e., overlapping) of data points that can naturally occur when reducing dimensionality. These benefits have been observed in recent architectural work that visualized categories
of GPU traffic patterns, with t-SNE providing substantial improvements in clustering behavior
compared to PCA [33]. Regardless, the standard t-SNE algorithm is also somewhat limited due
to its non-parametric nature, meaning that there is no "learned" transformation such as with the
eigenvectors in PCA. Consequently, online architectural applications are not as practical.

2.2.2   THE LEARNING PROCESS

We now describe a brief architectural example to further motivate unsupervised applications.
Consider a scenario in which we have many workloads and want to determine individual workloads that exhibit similar runtime behavior. As a general first step, we can profile workloads
to obtain feature vectors for resource usage, branch behavior, etc. Human-based classification
quickly becomes intractable if we wish to account for fine-grained, phase- or kernel-level behavior, since a single workload may have hundreds or thousands of fine-grained segments [33].
Supervised learning is, likewise, not possible in this scenario since we cannot train a model to
predict workload classes without already having examples for those classes. Instead, we can consider unsupervised learning methods. In the following, we exemplify k-means clustering as a
practical solution for such an architectural task.

The standard k-means clustering algorithm, shown in Algorithm 2.3, begins by randomly
initializing cluster centers (i.e., centroids). After that, an iterative clustering process cycles between assigning data points to the closest centroid and moving centroids to better represent the
points assigned to each cluster. In the context of our example scenario, each workload represents
one data point and workload classes represent clusters. Notice that this setup does make some
assumptions about the number of workload classes (the total number of initialized centroids).
In practice, however, this algorithm is quick enough that many values can be evaluated. Later
interpretation as labels for a supervised learning task may also dictate an appropriate number of
centroids.

Relatively simple modifications to Algorithm 2.3 can enable practical applications beyond
just data processing. One example can be found in related work on electronic design automation
(EDA). In that work [48], k-means clustering is used to reduce total wirelength by grouping
flip-flops into local regions where computation occurs. This clustering behavior is guided by
constraints on the number of flip-flops per cluster as well as the maximum flip-flop displacement,
essentially indicating when additional clusters are necessary. The generated designs reduced total
wirelength by 3.2% and total switching power by 4.8% compared to the prior state-of-the-art
approach.

Algorithm 2.3 Pseudocode for K-means clustering algorithm.

Input : Set of n data points $\mathcal{X} = \{x_1, \ldots, x_n\}$
Input : Set of m cluster centroids $\mathcal{K} = \{k_1, \ldots, k_m\}$

1: Randomly initialize cluster centroids
2: **while** Data points are changing clusters **do**
3:     **for all** Data points **do**
4:         Calculate the nearest centroid
5:         Assign the data point to that cluster
6:     **end for**
7:     **for all** Clusters **do**
8:         Calculate centroid location as the mean of all points assigned to that cluster
9:     **end for**
10: **end while**

2.3   SEMI-SUPERVISED LEARNING

Semi-supervised learning represents a mix of supervised and unsupervised methods, with some paired input/output data, and some unpaired input data. Using this learning approach, we could use unpaired data to perform clustering and then train a supervised model to determine appropriate cluster labels. Alternatively, we could train simple supervised learning models with the available labels, and then generate labels for the remaining data. We note that this approach has not yet found application in architecture. Nevertheless, work in other areas, including circuits analysis [49], provide some insight into potential applications.

In that work [49], semi-supervised learning is used to supplement limited supervised data and reduce required simulation time via Bayesian co-learning [50]. The main challenge observed in prior work is that in-depth circuit simulation and performance modeling has become impractical due to increases in circuit complexity and size. The solution, as proposed by Alawieh et al. [49], is to decompose the overall circuit under test into a combination of many distinct sub-circuits, each with a well-defined performance metric. In contrast to the overall design, these sub-circuits exhibit sufficiently low complexity that they can each be accurately modeled with a practical number of samples. These sub-circuit models, in turn, can be used to predict/generate labels for new data points, again at the sub-circuit level. These new data points, referred to as pseudo-samples, along with the original labeled data, provides a greatly expanded training set for the overall circuit model by recombining performance estimates from all the sub-circuits. Future work in architecture could explore a potential parallel with the hierarchical composition of architectural components.

2.4   REINFORCEMENT LEARNING

The supervised, semi-supervised, and unsupervised learning approaches in the previous sections assume that there exists a straightforward method to collect representative data without in-depth knowledge of the data itself. Standard workload profiling is one such method. Reinforcement learning (RL) instead addresses a distinctly different scenario in which data collection is a non-trivial process, particularly for time-sensitive tasks, and we begin with no data. Specifically, an RL model must generate its own data throughout the learning process.

> The need for representative data (i.e., broad exploration in reinforcement learning) is present throughout AI and learning algorithms due to the desire for highly generalizable models. In a supervised setting, we generally want a performance prediction model, for example, to make accurate predictions for as many possible workloads as possible. Similarly, in a reinforcement setting, the agent should explore sufficient states to ensure that estimates for rewards are valid. Representative datasets also help to mitigate the degree of incorrect predictions/actions that could negatively impact the system [51].

RL, as a whole, can be understood using a representation based on states, actions, and rewards, as seen in Figure 2.8. We elaborate on these concepts in the context of a memory controller. First is the environment, which represents the task that we seek to optimize, thus corresponding to the overall memory sub-system comprising incoming requests and outgoing responses. Within this task, the state represents the observable information about our environment. This could include the current number of queued read/write requests, the current activated row, etc. The agent is the decision-maker, continuously exploring the environment and gathering data by taking actions. These actions could include standard memory controller commands (e.g., row read, write, pre-charge, etc.). Finally, we have the rewards or a reward function, which guide the agent. The agent seeks to maximize its long-term rewards, so attempts to determine actions that, in each state, will maximize this reward. Memory throughput, for example, could be directly optimized by granting higher rewards for actions leading to high throughput. The agent's action-making strategy, also known as its "policy," is complicated by the need to explore alternative strategies that may lead to higher rewards.

2.4.1   MATHEMATICAL FOUNDATION

Most RL models share the same mathematical foundation and simply differ in their learning approach (i.e., the function or values that they attempt to estimate). We therefore begin with the common mathematical foundation and then move on to specific models in Section 2.4.2.

The mathematical foundation for RL is the Markov decision process. Formally, we represent a Markov decision process as

$$M = (S, A, R, P), \tag{2.30}$$

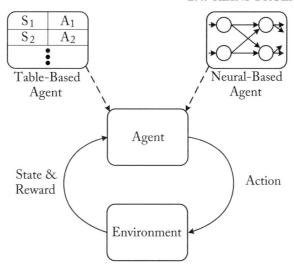

Figure 2.8: Generic reinforcement learning setup.

where S is the set of states, A is the set of actions, R defines the reward following a transition from state s to s', and P is the state-transition probability distribution $P(s'|s,a)$. This notation for the state-transition probability indicates that there are a number of possible, distinct states s' that can result from a specific action a in the current state s. In general, we can enumerate possible elements of S and A given knowledge of the environment. The difficulty in RL stems from limited knowledge about P and R. Again using our memory controller example, it is difficult to predict from the start how well various actions will improve long-term throughput, especially given uncertainty about future memory requests.

The overall goal of the RL agent is to maximize its long-term rewards or, equivalently, find a policy π that results in optimal rewards at each step. The value of a policy is then defined as

$$V^{\pi}(s) = \mathbb{E}_{\pi}[\sum_{t \geq 0} \gamma^t r_t | s_0 = s, \pi],\qquad(2.31)$$

where r_t is the reward at time t following each state transition and γ is a discount factor ($\leq$ 1), which dictates how much the model should consider future rewards; values close to one encourage the model to consider the long-term impact of its actions while values close to zero encourage the model to consider actions that maximize the immediate rewards, regardless of the consequences. The cumulative rewards are then maximized by an optimal policy π^* that satisfies

$$\pi^*(s) = \arg\max_{\pi} \mathbb{E}[\sum_{t \geq 0} \gamma^t r_t | s_0 = s, \pi].\qquad(2.32)$$

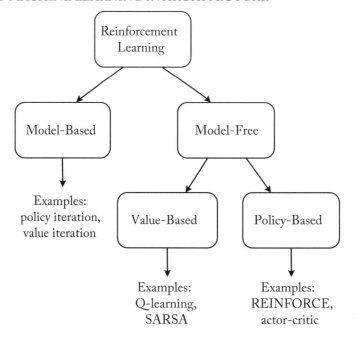

Figure 2.9: Hierarchy of reinforcement learning approaches.

2.4.2   MODELS

We illustrate a hierarchy of RL approaches in Figure 2.9. In general, the primary distinction between these approaches is the function/values that they learn to maximize long-term rewards. *Model-based* RL focuses on modeling P and R, essentially learning how the environment reacts to actions. These models can then be used to evaluate Equation (2.31) or Equation (2.32), thus serving as the basis for value iteration or policy iteration, respectively. In contrast, *model-free* RL methods attempt to model the value function $V^\pi(s)$ or the policy π directly.

In general, model-based RL tends to be more sample efficient [52]; samples can be used to update both P and R, as well as the derived policy/value estimates, while model-free RL updates just the policy/value function. Regardless, model-based RL may not always be practical due to calculation overhead. Observing Equation (2.32), we must still solve for an optimal policy, which involves a search that is not computationally efficient in a large state space.

RL application to architecture has, thus far, heavily favored model-free techniques. Within that category, there are two general implementation approaches. The first, *value-based* methods, approximate the values of each state and selects actions based on those with the highest estimated value. These methods, and Q-learning [53] in particular, have been the most common in architectural applications. Nevertheless, recent works have shifted toward the second category of *policy-based* methods, which seek to directly approximate the optimal policy. It is important to recognize that these various RL techniques modify the agent (upper portion of Figure 2.8), while the environment (i.e., the task) remains unchanged.

Value-Based Methods

At the foundation of many value-based methods is temporal-difference learning. The general idea in temporal difference learning is that, since we cannot directly evaluate Equation (2.31), we can iteratively update an estimate for the value function using

$$\hat{V}^{\pi}(s) = r_t + \gamma \hat{V}^{\pi}(s'). \tag{2.33}$$

Nevertheless, referring back to Equation (2.32), we still cannot determine an appropriate policy since we generally do not know the state transition probabilities, which are required to calculate the expected value over possible states. This limitation is especially true in large state spaces where it becomes impractical to model all possible state transition probabilities. Q-learning [53] solves this problem by replacing the value function with a Q-function $Q(s, a)$. This Q-function is similar to the original value function, but is defined over state-action pairs. Using this Q-function, we can observe quantitative differences between different actions in a given state. Viewed another way, the best action can be selected after evaluating all possible actions (or a subset due to resource/time constraints), without the need to model any state transition probabilities. Finally, a modified temporal-difference update for Q-learning is given as

$$Q(s, a) = (1 - \alpha)Q(s, a) + \alpha[r(t) + \gamma \max_{a' \in A} Q(s', a')]. \tag{2.34}$$

Notice that this update uses the maximum achievable Q-value in the next state s' even though this optimal action might not actually be selected.[6] Q-learning is therefore called an "off-policy" algorithm since learning does not strictly follow taken actions. In contrast, an "on-policy" algorithm such as SARSA [54] always considers the actual actions taken, resulting in an update given as

$$Q(s, a) = (1 - \alpha)Q(s, a) + \alpha[r(t) + \gamma Q(s', a')]. \tag{2.35}$$

Additionally, observe that the Q-function update involves a hyperparameter α that acts as the learning rate for the function, essentially smoothing updates to maintain a running average rather than dramatically changing after each update.

Policy-Based Methods

Policy-based methods take a different approach from value-based methods by optimizing the policy directly, thus eliminating the need to model either the environment $P(s, s')$ or the value function $V(s)$. Previously, the policy would be represented as either $\pi(s)$ for a deterministic

---

[6]It is common practice for RL models to randomly select new states/actions for exploration, especially during the start of training. This exploration helps ensure that the learned policy is eventually optimal.

policy or $\pi(s|a)$ for a stochastic policy.[7] Here, in policy-based methods, a stochastic policy is represented as $\pi(a|s, \theta)$ where θ is included as the parameter to be learned. The idea is that we can optimize θ with respect to the gradient of some performance measure (i.e., $\nabla J(\theta)$) using gradient ascent. This approach is the basis for the REINFORCE [56] algorithm. Other methods exploit additional information: actor-critic methods [55] also model the value function as a more reliable baseline for the benefit of a particular action vs. the current policy. Notably, these stochastic gradient-based learning approaches can be directly applied to continuous action spaces.

Model Implementation

Both the model-free and model-based learning approaches described in previous sections can be implemented a number of different ways. For example, Q-learning was traditionally implemented using a tabular approach with rows and columns for all possible states and actions in those states. In this setup, each entry in the table corresponds to a Q-value estimate. The recent deep learning paradigm has prompted new approaches that synthesize neural networks and reinforcement learning algorithms, resulting in DRL. In this approach, a DNN is used to approximate RL functions, such as the Q-function in Q-learning, or the environment in model-based RL. A detailed DRL approach is covered as a case study in Chapter 4.

2.4.3   THE LEARNING PROCESS

The standard Q-learning algorithm is shown in Algorithm 2.4. There are a few particular aspects to observe. First, note that the action selection (line 5) based on policy Q is not necessarily the action that maximizes $Q(s, a)$. This is because the agent must try alternative, previously unexplored actions to ensure that the Q-function is representative of all possible state-action pairs. Second, the Q-function update in line 6 is only necessary during an initial training phase, assuming the task does not require online updates.

2.5   EVALUATION METRICS

Continued research involving machine learning relies upon a variety of metrics to ensure fair comparison between works. Although these machine learning metrics usually remain applicable in architectural applications, improvements in machine learning metrics can adversely impact more traditional architecture metrics. Consider, for example, a complex regression task that involves nonlinear relationships between many features. In this scenario, any linear model would fail to capture some relationships and, in general, exhibit relatively high error. Alternatively, we would expect that a more complex model (e.g., an ANN with multiple layers) could more accurately model any complex relationships, resulting in lower prediction error. The problem,

---

[7]Briefly, deterministic policies always take the same action in a given state (assuming no policy updates) while stochastic policies sample from a distribution of actions in a given state. We encourage the interested reader to check Sutton [55] for further discussion.

Algorithm 2.4 Pseudocode for Q-learning algorithm.

Input: Set of valid actions A
Output: Q-function $Q(s, a)$, an estimator of Q-value for all $a \in A$

1: Initialize $Q(s, a)$ arbitrarily, $Q(terminal\_state) = 0$
2: **while** Terminal condition has not been reached **do**
3:   Set s as the current state
4:   Choose action a for the current state s based on policy Q
5:   Take action a and observe the resulting state s' and reward r
6:   Update $Q(s, a) = (1 - \alpha)Q(s, a) + \alpha[r + \gamma \max_{a' \in A} Q(s', a')]$
7: **end while**

however, is that the higher complexity model will require more energy, time, and/or resources to generate a single prediction. In some architectural applications, this overhead may even outweigh the benefits of more accurate predictions. As a result, machine learning application to architecture tends to exhibit complex trade-offs that require careful consideration for practical implementation. Here, we introduce a few prevalent metrics as a primer for trade-offs highlighted in the following chapters. Many readers may already be familiar with these metrics, but readers from other backgrounds may still benefit from this discussion.

Machine learning metrics can generally be categorized along with the learning method or task.[8] In the case of classification tasks, model performance is commonly evaluated with a *confusion matrix*, with an example two-class[9] matrix illustrated as Table 2.1. This table counts the number of instances (i.e., data points) that are predicted for each class compared to their actual class. Several metrics can be calculated from this table:

- Precision, which is the fraction of predicted positive instances that are actually positive, is calculated as

$$Precision = \frac{TP}{TP + FP}. \tag{2.36}$$

- Recall (also known as sensitivity), which is the fraction of actually positive instances that are predicted positive, is calculated as

$$Recall = \frac{TP}{TP + FN}. \tag{2.37}$$

[8]Machine learning metrics for reinforcement and unsupervised learning are generally more task-specific so are presented as necessary in the case studies in Chapter 4.
[9]Confusion matrices and its metrics can also be extended to multi-class problems.

Table 2.1: Confusion matrix

		Predicted Class	
		Positive	**Negative**
Actual	**Positive**	True Positive (TP)	False Negative (FN)
Class	**Negative**	False Positive (FP)	True Negative (TN)

- Specificity, which is the fraction of actually negative instances that are predicted negative, is calculated as

$$Specificity = \frac{TN}{TN + FP}. \tag{2.38}$$

- Accuracy, which is the fraction of all predictions that are correct, is calculated as

$$Accuracy = \frac{TP + TN}{TP + TN + FP + FN}. \tag{2.39}$$

For regression tasks, we have already observed one example metric, MSE. Another similar metric is the mean absolute error (MAE), which instead uses the magnitude of the error (i.e., L1 norm), given as

$$MAE = \frac{1}{N} \sum_{i=1}^{N} |y_i - \hat{y}_i|. \tag{2.40}$$

All the aforementioned machine learning metrics focus on how well a model either distinguishes particular examples or predicts values that are similar to training examples. Improvements in these metrics ultimately require models with sufficient learning capabilities or, in other words, complexity. This complexity, in turn, means that more data must be collected for model input, more storage is required for model weights, and more calculations are required to produce each prediction. Specifically, machine learning metrics may be at odds with traditional architecture metrics including the following.

- Execution time measures the total time for an application (or set of applications) to complete its task. High-complexity machine learning models can increase execution time if hardware resources must be shared or are otherwise impacted by machine learning data collection and prediction.

- Area refers to the total physical space required for hardware components. Dedicated machine learning hardware can improve execution time, but increase area overhead.

- Power usually refers to the average power consumption of a system. Modern systems are typically power-limited (due to thermal constraints), thus any power required to generate predictions may reduce power available for other components.

- Energy is measured as the average power multiplied by execution time. Similarly, energy efficiency is the amount of useful calculations divided by energy required to perform those calculations. Simpler machine learning models tend to require fewer calculations, thus incurring less energy overhead. These simple models, however, may not predict optimal system configurations, resulting in overall lower energy efficiency for longer tasks.

- Task-specific metrics may also be negatively impacted. For example, work on cache prefetching usually considers coverage, which is the fraction of cache misses eliminated by prefetching, timeliness, which measures whether prefetched data arrived when it was needed by the processor, and prefetch accuracy, which is the fraction of prefetches that were actually used by the processor. These works may favor more complex predictors since inaccurate predictions/prefetches may evict useful entries and reduce overall performance. Of course, this complexity can negatively impact other metrics (e.g., area).

- Throughput is another task-specific performance metric. Specifically, throughput may refer to the total tasks processed in unit time (at a system level) or total packets processed in unit time (in NoCs) or memory requests processed in unit time (for memory systems). This metric generally aligns better with machine learning metrics, but, as mentioned above, better predictions may require more complex models and higher overhead, particularly for area and power.

Optimizing these architectural metrics can be challenging even with traditional design methods. Additional consideration for machine learning metrics therefore introduces substantial complexity, leading to diverse possibilities for architectural applications.

SUMMARY

The machine learning approaches discussed in this chapter offer distinct capabilities that can be directly applied to some of the most challenging architectural problems. Supervised learning provides a reliable method to approximate practically any function, enabling tasks such as performance and power prediction as well as proactive and dynamic control of workloads and resources. Unsupervised learning can perform automatic data analysis and clustering without human intervention, thus uncovering new strategies or information that can guide future designs. Semi-supervised learning, although limited in current applications, represents a foundation to expand supervised learning principles into domains with limited data availability, potentially playing a critical role as architectural complexity continues to increase. Finally, RL facilitates intelligent exploration in expansive design spaces such as those in architecture, resulting in practical design optimization that can, again, scale with architecture complexity. Note that the models presented in this chapter represent just a small fraction of the existing implementation possibilities. We will see many more examples of these exciting new capabilities and applications in Chapter 3.

CHAPTER 3

Literature Review

The previous chapter on machine learning fundamentals provided a brief glimpse into the diverse applications for machine learning in architecture. Now, in this chapter, we review current literature to explore the full breadth of these applications and their already significant contributions to the state-of-the-art. Each work is, out of necessity, only discussed briefly as we highlight any significant aspects of the problem setup, implementation, or particularly noteworthy results. For that reason, the primary goal for the readers in this chapter is to continue building intuition about how machine learning can be applied. We will apply this intuition to the case studies in Chapter 4.

Research is organized by sub-systems, when applicable, or primary objectives. We begin with work applying machine learning to general system simulation and performance/power prediction. We then consider applications in several specific sub-systems include GPUs, memory and branch prediction, and the NoC. After that, we examine system-level optimization and approximate computing, which integrate a variety of techniques across sub-systems. Following each section, summary tables are provided that group works by their goal. Works with identical goals are denoted by a ditto mark, ", following the first entry.

3.1   SYSTEM SIMULATION

Cycle-accurate full-system simulations are commonly used in system performance modeling, but require several orders of magnitude more time than native execution. Machine learning methods can help to offset this penalty, either by improving more traditional simulators or by directly predicting performance, thus offering a trade-off between simulation time and accuracy. In general, machine learning can reduce execution time by 2–3 orders of magnitude with relatively high accuracy (task dependent, typically > 90%).

Several existing simulation frameworks incorporate machine learning methods to improve efficiency or extend simulator functionality. SimPoint [57] uses k-means clustering to identify application phases (i.e, intervals with distinct behavior), thus enabling simulations based on a small, yet representative subset of instructions. Newer SimPoint versions additionally provide support for variable phase lengths using a weighted k-means scheme. SynFull [58], a NoC simulation tool proposed by Badr and Enright Jerger, uses statistical analysis to generate realistic communication behavior with reduced simulation overhead. Representative long-term traffic behavior is determined using a variant of k-medoids[1] and then controlled with a Markov chain. Later work (APU-SynFull) [59] expanded simulation capabilities by generating realistic communication behavior for arbitrary NoC sizes (as well as heterogeneous architectures) via regression modeling.

An alternative approach used in several early works directly predicts system performance using trained machine learning regression models. For example, Ipek et al. [60] modeled an architectural design space using an artificial neural network ensemble (i.e., a group of ANN predictors). Models were trained on approximately 1% of the design space, then predicted chip multiprocessor performance with 4–5% error for random points, albeit only in that specific configuration space. When combined with SimPoints, predictions exhibit slightly higher error, but the simulated instruction count is further reduced. Ozisikyilmaz et al. [61] additionally predicted SPEC performance for future systems that may be poorly modeled by existing simulators. Evaluation was limited to randomly sampled data with relatively simple linear regression and neural network models, yet demonstrated advantages compared to single-layer models (as in [60]).

Machine learning approaches have also been proposed for more general system modeling. Eyerman et al. [23] proposed a mechanistic-empirical model for processor cycles-per-instruction (CPI) prediction. In this approach, they used a generic mechanistic model[2] with parameters inferred by a regression model. Their model is limited to single-core performance prediction, but provides several key improvements: both accuracy and ease of implementation are improved compared to a purely mechanistic model while interpretability is improved compared to a purely empirical model. Zheng et al. [39, 62] explored cross-platform application performance prediction for ARM processors based on measurements from Intel and AMD processors. Their first approach [39] using linear regression predicted performance based on a local neighborhood of examples around the target point to approximate nonlinear behavior. They later [62] emphasized phase-level performance prediction, assuming that phase-level behavior would be approximately linear. Notably, the average error for cycle count predictions is less than 1% using phase-level profiling. This approach is, however, restricted to a single target architecture and requires source code for phase-level analysis, leaving significant opportunities for future work. Finally, recent work by Agarwal et al. [63] introduced a method to predict parallel execution speedup using single-threaded execution characteristics. They trained separate models for each thread count using application-level performance counters. Although neural networks were omitted due to limited data, evaluation found that Gaussian process regression still provided promising results, particularly for high thread counts.

---

[1]At a high level, k-medoids differs from k-means in that k-medoids uses an appropriate data point as the cluster identifier while k-means uses the mean of all data points in a cluster.

[2]Here, mechanistic refers to a model based on underlying mechanisms, e.g., the amount of instructions that can be processed in parallel.

Table 3.1: Summary of works on machine learning for system simulation

Work	Goal	Strategy	Lessons Learned
[57]	Determine representative workload behavior	K-means on basic block frequency vectors	Consider dataset reduction prior to clustering and weigh data points when appropriate
[58]	" "	K-medoids on tracffic flow vectors	Test multiple strategies when selecting features and model parameters
[59]	Predict performance with scaled NoC resources	Linear regression on easily simulatable configurations	Consider splitting problems into multiple simpler models
[60]	Predict performance in CPU design space	ANN ensemble regression on small subset of points	Explore complementary techniques from machine learning and traditional methods
[61]	Predict performance for future systems	Linear/ANN regression on common processor parameters	Complex models can perform worse on comparatively simple tasks
[23]	Predict and interpret processor performance	Nonlinear regression on hard-to-determine performance indicators	Apply domain knowledge when possible to guide machine-learning models
[39, 62]	Cross-platform CPU performance prediction	Linear regression on common performance counters	Exploit similarities in phase-level behavior
[63]	Predict performance scaling with thread count	Regression with several models on common performance counters	Recognize limitations in data availability when training complex models

3.2   GPUS

3.2.1   DESIGN SPACE EXPLORATION

GPU design space exploration has proven to be a particularly favorable application for machine learning due to its highly irregular design space; some kernels exhibit relatively linear scaling while others exhibit very complex relationships between configuration parameters, power, and performance [46, 64, 65].

Jia et al. [64] proposed Stargazer, a regression-based framework implemented using natural cubic splines. Stargazer randomly samples approximately 300 points from a target design space (933 K points in evaluation) for each application, then applies stepwise regression on these points. Notably, the framework achieves under 3.8% average performance prediction error. Similar work by Jooya et al. [65] considered a per-application performance/power prediction model, but additionally proposed a scheme to predict per-application Pareto fronts. Many ANN-based predictors were trained and the most accurate subset was used as an ensemble for prediction. Prediction accuracy was later improved by sampling points within a threshold of the previously predicted Pareto-optimal curve. Wu et al. [46] instead explicitly modeled scaling for compute units, core frequency, and memory frequency. Scaling data from training kernels was processed using k-means clustering to group kernels by scaling behaviors. An ANN then classifies the kernels into these clusters, allowing new kernels to be classified and predictions made using cluster scaling factors. This approach, in contrast to Jia et al. [64], requires just a few samples for new applications.

Lin et al. [66] combined a performance-predicting DNN with a genetic search scheme to explore memory controller placement. The DNN was used as a surrogate fitness function, obviating slow system simulations. The resulting placement improves system performance by 19.3%.

3.2.2   CROSS-PLATFORM PREDICTION

Porting applications for execution on GPUs is a challenging task with potentially uncertain benefits over CPU execution. Work has therefore examined methods to predict speedup or efficiency improvements using just CPU execution behaviors.

Baldini et al. [29] cast the problem as a classification task, training a modified nearest-neighbor and an SVM model to determine, based on a threshold, whether GPU implementation would be beneficial. Using this approach, they predicted near-optimal configurations 91% of the time. In contrast, Ardalani et al. [67] trained a large ensemble of regression models to directly predict GPU performance for the code segment. Although several code segments exhibit high error, the geometric mean of the absolute value of the relative error is still just 11.6% and the model successfully identifies several code segments that are incorrectly predicted by human experts. Later work by Ardalani et al. [68] introduced a completely static-analysis-based framework using a random forest model for binary classification. This approach eliminates both dynamic profiling and human guidance, instead using features such as instruction mix, branch divergence estimates, and kernel size to provide 94% accuracy for binary speedup classification with a speedup threshold of 3.

3.2.3   GPU SPECIFIC PREDICTION AND CLASSIFICATION

O'Neal et al. [69] presented a methodology for next-generation GPU performance prediction as cycles-per-frame (CPF) for DirectX applications. They focused on Intel GPUs, profiling earlier-generation architectures (e.g., Haswell GT2) to train next-generation predictors. They found that prediction accuracy for linear and nonlinear models is dependent upon the target architecture, with the best performing models achieving less than 10% CPF prediction error.

Recent work by Li et al. [33] presented a re-evaluation of commonly accepted knowledge of GPU traffic patterns. They used a CNN and t-distributed stochastic neighbor embedding (t-SNE) on heatmap-transformed traffic data, identifying eight unique patterns with 94% accuracy.

3.2.4   SCHEDULING

GPU processing-in-memory (PIM) architectures can benefit from high memory bandwidth and reduced data movement energy. Despite these benefits, potential limitations on PIM compute capabilities may introduce complex trade-offs between performance and energy when scheduling execution on various resources. For this reason, Pattnaik et al. [70] proposed an approach using a regression model to classify core affinity, thus dividing the workload, and an additional regression model to predict execution time, enabling dynamic task migration. Performance and energy efficiency are improved by 42% and 27%, respectively, over a baseline GPU architecture. Future work could explore alternative models for core affinity classification to better predict runtime-dependent behavior, thereby further improving performance and/or energy efficiency.

3.3   MEMORY SYSTEMS AND BRANCH PREDICTION

3.3.1   CACHES AND PREFETCHING

Traditional approaches for caching can incur performance penalties due to dramatic workload variance. In contrast, machine learning approaches can learn these intricacies and offer better performance.

Peled et al. [71] proposed a prefetcher exploiting semantic locality (i.e., locality derived from data structures) using contextual bandits (a simple RL variant), correlating contextual information and candidate addresses for prefetching. Implementation uses a two-level indexing method to dynamically control state information, allowing online feature selection with some additional overhead. Zeng and Guo [72] proposed an LSTM model for regression-based prefetching using local history and offset-delta tables. Evaluation showed that the LSTM model enables accurate predictions over longer sequence and higher noise resistance than prior work, albeit on synthetic traces. Later work by Hashemi et al. [73] instead used a classification-based approach for their LSTM model, noting improved precision/recall on irregular patterns and real workloads. In addition, they explored a clustering approach for the address space, allowing the LSTM model to better adapt to local context. Similarly, Braun et al. [74] extensively explored LSTM prefetching accuracy under several common access patterns. Experiments considered the impact of lookback size (access history window) and LSTM model size for several noise levels and independent access stream counts. Recent work by Bhatia et al. [75] synthesized traditional prefetchers with a perceptron-based prefetch filter, allowing aggressive predictions without degrading accuracy. Evaluation confirmed substantial coverage and IPC benefits offered by the proposed scheme, with 9.7% IPC speedup over the next best prefetcher when referenced to a no-prefetching four-core baseline. Increasingly multi-core designs have also necessitated more

Table 3.2: Summary of works on machine learning for GPUs

Work	Goal	Strategy	Lessons Learned
[64]	Predict performance on GPU design space	Nonlinear regression on common GPU parameters	Take time to investigate unexpected behaviors in machine-learning models
[65]	" "	ANN ensemble regression on common GPU parameters	Focus learning on near-optimal region
[46]	Predict performance with scaled runtime resources	K-means on performance/power scaling, neural net classication on GPU performance counters	Explore clustering on individual performance factors, rather than all together
[66]	Optimize GPU memory controller placement	Genetic search with DNN-based fitness function	Reduce evaluation time using simpler machine-learning-based predictions
[29]	Predict GPU speedup from CPU runtime execution statistics	Multiple classification models on instruction mix counters	Explore various bins and thresholds for classifications tasks
[67]	" "	Ensemble of nonlinear regression models on instruction, cache, and memory performance counters	Apply ensembles to mitigate the impact of poor predictors
[68]	Predict GPU speedup from static code-based features	Random forest classification on instruction frequency and memory/branch behavior	Apply domain knowledge to guide feature selection
[69]	Predict cycles-per-frame in DirectX applications	Ensemble of linear/nonlinear regression models on common performance counters	Simpler (linear) models can provide better performance in some tasks
[33]	Identify GPU traffic patterns	Convolutional neural network on heatmap transformed traffic flow data, then clustered by t-SNE	Consider convolutional neural networks to provide alternative representations for matrix-formatted data
[70]	Schedule tasks on heterogeneous GPU architecture	Logistic/linear regression for affinity/execution time prediction on common performance indicators	Consider splitting problems into multiple simpler models

holistic consideration for prefetching and contention. Specifically, Hiebel et al. [76] explored the impact of dynamic prefetcher configuration (i.e., enabling/disabling) on a per-core basis to mitigate harmful memory bandwidth contention. Using a contextual bandits model, they observe up to 4.3% average speedup over a static configuration with all prefetchers enabled.

Machine learning has similarly been applied to data reuse policies. For example, Teran et al. [77] predicted LLC reuse with a perceptron model. In this approach, input features are hashed to access saturating weight tables that are incremented/decremented based on correct/incorrect reuse prediction. These features are chosen empirically and shown to significantly impact performance, thus presenting an option for further optimization. Wang et al. [78] likewise predicted reuse prior to cache entry, only storing data in the cache if there was predicted reuse. They used decision trees as a low-cost alternative to ensemble models, achieving 60–80% reduction in writes.

Additional research has explored the growing performance bottleneck in translation lookaside buffers (TLBs). In particular, Margaritov et al. [79] proposed a scheme for virtual address translation in TLBs based on learned indices [80]. Evaluation showed nearly 100% accuracy for predicted indices, but practical implementation will require dedicated hardware to reduce calculation overhead (and is left for future work).

3.3.2   SCHEDULERS AND CONTROL

Controllers for memory and storage systems influence both device performance and reliability, thus representing another strong application for machine learning models compared with heuristics.

Ipek et al. [15] first proposed an RL approach for memory controllers to capture the balance between concurrency, delay, and several other factors. The model predicted optimal actions (e.g., precharge, activate, row read/write), improving system performance by 15% over prior work. Mukundan and Martinez [81] proposed improvements by generalizing the reward function, enabling diverse optimizations goals for energy, fairness, etc. They also added power-up and power-down actions to enable a further 8.6% improvement in performance and a significant improvement in energy efficiency.

Related work also optimizes communication energy between memory/storage and other systems using machine learning. Manoj et al. [82] proposed a Q-learning method for dynamic voltage control in through-silicon-interposer transmission lines. Predictions for power and bit error rate are quantized, then provided as input to the model to determine a new voltage level. Although their approach requires significant quantization to minimize overhead, they still achieved 15.1% energy savings compared to a static voltage baseline. Wang and Ipek [83] reduced data movement energy through online clustering and encoding. Several clusters are continuously updated at a bit-level using majority voting for data in that cluster. The total number of transmitted 1s is then minimized by XORing new data with the closest learned cluster center.

Kang and Yoo [84] applied Q-learning to manage garbage collection in SSDs by determining optimal periods of inactivity. Key states are kept in the Q-table using LRU replacement, allowing a vast state space and, ultimately, a 22% average tail latency reduction over the baseline. Many states are, however, observed only once per workload, suggesting potential benefits using deep Q-learning.

Other work directly considered system reliability. For example, Deng et al. [85] proposed a regression-based framework to dynamically optimize performance and lifetime in non-volatile memories (NVMs). Their approach used phase-based application statistics to manage several conflicting policies for write latency, write cancellation, endurance, etc., guaranteeing a minimum lifetime with modest performance/energy improvements. Xiao et al. [86] proposed a method for disk failure prediction using an online random forest. They trained their model using a disk status window to account for imprecision in recorded failure date, enabling accurate predictions of soon-to-be faulty drives. Comparison against other random forest updating schemes (e.g., updating once a month) highlighted accuracy benefits from consistent training that may be extended to related domains.

3.3.3   BRANCH PREDICTION

Branch prediction is a noteworthy example of current machine learning application in industry, with accuracy surpassing prior state-of-the-art non-machine-learning predictors. The perceptron-based branch predictor was first proposed by Jiménez and Lin [14] as a promising high-accuracy alternative to two-level schemes using pattern history tables. Later research by St. Amant et al. introduced SNAP [87], a perceptron-based predictor implemented using analog circuitry to enable an efficient and practically feasible design. Perceptron weights and branch history were used to drive current-steering digital-to-analog converters that perform the dot product as the sum of currents. Jiménez [88] further optimized this design using a per-branch history table, dynamic coefficients for history importance, and a dynamic learning threshold. The optimized design achieves 3.1% lower misses per kilo-instructions (MPKI) than L-TAGE [21]. Recent work with perceptron-based predictors by Garza et al. [89] explored bit-level prediction for indirect branches. Possible branch targets are evaluated via dot product with stored weights from eight feature tables incorporating local and global history. The resulting predictions achieve state-of-the-art accuracy and reduce MPKI by 5% compared to ITTAGE [90].

Current state-of-the-art conditional branch predictors (e.g., TAGE-SC-L [19]) are prone to systematic mispredictions for a small number of static, hard-to-predict branches (H2Ps) that hide significant IPC gains of 14.0% for an Intel Skylake architecture [91]. Tarsa et al. [91] consequently proposed "CNN Helper" predictors that target specific H2Ps using simple two-layer CNNs. Results indicate strong applicability across diverse workloads and present a promising area for future work.

Table 3.3: Summary of works on machine learning for memory systems and branch prediction (*Continues*)

Work	Goal	Strategy	Lessons Learned
[71]	Develop prefetcher to exploit locality in data structures	Contextual bandits on recent instruction and branch history, reward based on usefulness	Consider experimenting with dynamic feature selection to reduce overfitting or improve accuracy
[72]	Explore prefetcher for long sequences	LSTM regression using local history and offset-delta history	Machine learning models can provide high noise tolerance compared to conventional methods
[73]	" "	LSTM classification on program counter and local address-delta, k-means on address-delta groups	Memory deltas are easier to learn than all possible distinct values
[74]	" "	LSTM classification on per-PC memory address deltas	Model parameters can significantly impact some tasks, while having little impact on others
[75]	Improve prefetch aggressiveness without harming accuracy	Traditional prefetcher with perceptron-based filter using common address features and XOR hashed features	Combining traditional and machine learning models can enable improved trade-offs in some tasks
[76]	Optimize multi-core prefetcher configuration	Contextual bandits on common performance counters	Per-core models can still cooperate using global performance counters
[77]	Predict LLC reuse for general data	Perceptron classification on PCs, block tag, and hashed versions	Shifted counter values can provide alternative perspectives for learning
[78]	Predict LLC reuse for image data	Decision tree classifier on average image views, type, size, age, etc.	Simple decision trees can provide appropriate performance in seemingly difficult tasks
[79]	Improve TLB address translation via index predictions	Hierarchy of ANN regressors using Linux page table dump	Practical applications necessitate trade-offs between overhead and accuracy

Table 3.3: (*Continued*) Summary of works on machine learning for memory systems and branch prediction

[15]	Explore machine-learning-based memory controllers	Q-learning using pending reads/writes and command under consideration	Exploit online updating when possible to better meet changing workload demands
[81]	" "	Q-learning similar to [15], additional actions to power/up down DRAM rank	Alternative reward structures provides simple method to balance diverse goals
[82]	Minimize data communication energy via voltage control	Q-learning using current power and bit error rate	Quantization offers a simple, yet effective method to trade overhead and accuracy
[83]	Minimize data communication energy via data encoding	K-means on data to be transmitted, XOR with identified cluster centers	Online updates can be made more efficient with longer intervals
[84]	Optimize SSD garbage collection	Q-learning on history of request interval, sizes, and actions	Online RL allows effective management of states to reduce storage
[85]	Optimize performance, energy, and reliability in NVMs	Several regression models using configuration options mapped as numerical values	Try to understand important features and their impact on predictions
[86]	Predict disk failure	Random forest classifier on SMART features	Remember to account for model bias when dataset is small/skewed
[14, 87, 88]	Explore perceptron-based branch prediction	Mixed digital-analog perceptron using per-branch history table	Simple perceptron models can provide state-of-the-art accuracy in complex tasks
[89]	Explore perceptron-based indirect branch prediction	Perceptron using common features and hashed version for each bit of target address	Task-specific optimizations can significantly improve performance
[91]	Predict specific hard-to-predict conditional branches	Convolutional neural network using encoded history of instruction pointer and branch direction	Even two-bit ANNs can achieve state-of-the-art accuracy when trained on highly specific targets

3.4   NETWORKS-ON-CHIP

3.4.1   DVFS AND LINK CONTROL

Modern computing systems exploit complex power control schemes to enable increasingly parallel architectural designs. Traditional schemes may fail to exploit all energy-saving opportunities, particularly in dynamic NoC workloads, leading to significant benefits through proactive machine-learning-based control.

Savva et al. [92] implemented dynamic link control using several ANNs, each of which monitors a NoC partition. These ANNs used just link utilization to learn a dynamic threshold to enable/disable links. Despite energy savings, their approach can cause high latency under dimension-ordered routing. DiTomaso et al. [93] relocated flit buffers to the links and dynamically controlled both link direction and power-gating with per-router classification trees. Using a simple three-level tree to limit overhead, overall NoC power is reduced by 85% and latency is reduced by 14% compared to a concentrated mesh. Winkle et al. [94] explored machine-learning-based power scaling in photonic interconnects. Even a simple linear regression model provides promising results, negligibly reducing throughput (vs. no power-gating) while reducing laser power consumption by 42%. Reza et al. [95] proposed a multi-level ANN control scheme that considered both power and thermal constraints on task allocation, link allocation, and node DVFS. Individual ANNs classify appropriate configurations for local NoC partitions while a global ANN classifies optimal overall resource allocation. This scheme identifies the global optimal NoC configuration with high accuracy (88%), but uses complex ANNs that could impact implementation. Clark et al. [96] proposed a router design for DVFS and evaluated several regression-based control strategies. Variants predict buffer utilization, change in buffer utilization, or a combined energy and throughput metric. This work was expanded by Fettes et al. [97], who introduced an RL control strategy. Both regression and RL models enable beneficial trade-offs, although the RL strategy is most flexible.

3.4.2   ADMISSION AND FLOW CONTROL

As with NoC DVFS, both admission and flow control can benefit from proactive prediction. Early work by Boyan and Littman [98] introduced Q-learning-based routing (Q-routing) for generalized networks using delivery time estimates from neighboring nodes, noting throughput advantages over traditional shortest path routing for high traffic intensity. Several works have expanded upon Q-routing, observing applications in dynamically changing NoC topologies [99], improved capabilities in bufferless NoC fault-tolerant routing [100], and high-performance congestion-aware non-minimal routing [101].

More recent works have focused on injection throttling, traffic arbitration, and hotspot prevention. For example, Daya et al. [102] proposed SCEPTER, a bufferless NoC using single-cycle multi-hop paths. They controlled injection throttling using Q-learning to maximize multi-hop performance and improve fairness by reducing contending flits. Future work could reduce

Q-table overhead which scales with NoC size in their implementation. Soteriou et al. [103] instead focused on injection throttling for hotspot prevention. Their ANN was first trained to recognize hotspot formation based on buffer utilization and then trained to recognize the impact of proposed injection throttling and dynamic routing, providing a holistic mitigation strategy. The model provides state-of-the-art results for throughput and latency under synthetic traffic, but limited improvement under real-world benchmarks, suggesting the potential for further optimization. Similar work by Wang et al. [104] used an ANN to predict optimal NoC injection rates. Additional preprocessing (to capture both spatial and temporal trends) and node grouping enables high accuracy predictions (90.2%) and reduces execution time by 17.8% compared to an unoptimized baseline. Yin et al. [105] explored a deep Q-learning (DQL) approach for dynamic traffic arbitration. They considered a wide range of features and rewards while noting that the proposed DQL algorithm is impractical due to overhead. Regardless, evaluation exhibited modest throughput improvements over round-robin arbitration. More recent work by Yin et al. [106] presented a detailed case study on ML-driven design for NoC arbitration. As with branch prediction, NoC arbitration mandates cycle-level decisions, thus hindering application with standard machine learning approaches. Instead, they derive a practical human-expert implementation using insights gained from a trained deep Q-learning model, ultimately providing a practical approach rivaling the original machine learning model.

3.4.3   TOPOLOGY AND GENERAL DESIGN

Several works have leveraged machine learning in higher-level NoC topology design involving trade-offs between power and performance, with some work further considering thermal limitations.

Das et al. [107] used a machine-learning-based *STAGE* algorithm to efficiently explore small-world inspired 3D NoC designs. In this approach, design alternates between local search (adding/removing links in a hill-climbing approach) and meta search (predicting beneficial starting points for local search using prior results). The same model was used again by Das et al. [108] to balance link utilization and address through-silicon-via reliability concerns. The STAGE algorithm was then enhanced by Joardar et al. [109] to optimize a heterogeneous 3D NoC design. The models explores multi-objective trade-offs between CPU latency, GPU throughput, and thermal/energy constraints. All three works still rely upon hill-climbing for optimization. Recent work by Lin et al. [17, 110] instead explored DRL in routerless NoC design. They used a Monte Carlo tree search to efficiently explore search spaces surpassing 10^{100} and a deep convolutional neural network to approximate both the action and policy functions, thereby optimizing loop configurations. Further, the proposed DRL framework can strictly enforce design constraints that may be violated by prior heuristic or evolutionary approaches.

Limited work has also considered substantially larger NoC and system-on-chip (SoC) design spaces. In particular, Rao et al. [111] investigated multi-objective optimization involving features such as NoC bandwidth requirements and SoC area. Machine learning models were trained using data from thousands of SoC configurations to predict optimal NoC designs based on performance, area, or both. Limited comparisons against human-expert designs did not consider alternative techniques (e.g., AMOSA [112]), yet exhibited some promising results, motivating research into effective features and models as well as further comparisons against alternative techniques.

3.4.4   PERFORMANCE PREDICTION

Existing NoC models based on queuing theory are generally accurate, but rely upon assumptions of traffic distribution that may not hold for real applications [113]. Qian et al. [113] emphasized how machine-learning-based approaches can relax the assumptions made by queueing theory models. They constructed a mechanistic-empirical model based on a communication graph, using support vector regression to relate several features and queuing delays. The resulting model exhibits 3% error, compared to 10% for an existing analytical approach. Sangaiah et al. [114] considered both NoC and memory configuration for performance prediction and design space exploration. Following a standard approach, they sampled a small portion of the design space, then trained a regression model to predict the resulting system CPI. Evaluation generally shows high accuracy, but lower accuracy for high-traffic workloads (median error of 24%). Additional design space exploration exhibits promising results, reducing the design space from 2.4 M points to less than 1000.

3.4.5   RELIABILITY AND ERROR CORRECTION

Overhead introduced by error correction in NoCs can be significant, especially when retransmission is required. Several works have therefore explored machine-learning-based control schemes that can proactively predict errors and adjust correction mechanisms.

DiTomaso et al. [18] trained a decision tree to predict NoC faults using a wide range of parameters including temperature, utilization, and device wear-out. These predictions allow proactive encoding (on top of the baseline cyclic redundancy check) for transmissions that are likely to have errors. Wang et al. [115] adopted a similar strategy for dynamic error mitigation, but used an RL-based control policy to eliminate the need for labeled training examples. Their approach provides average dynamic power savings of 46% compared with a static CRC scheme and 17% compared with the decision tree method [18]. Wang et al. [116] subsequently proposed a holistic framework for NoC design incorporating dynamic error mitigation, router power-gating, and multi-function adaptive channel buffers. They emphasized comprehensive benefits through synergistic integration/control of several architectural innovations, thus achieving substantial improvements in latency (32%), energy-efficiency (67%), and reliability (77% higher Mean Time to Failure) compared to a SECDED baseline.

Table 3.4: Summary of works on machine learning for NoCs (*Continues*)

Work	Goal	Strategy	Lessons Learned
[92]	Dynamic link control (enabled/disabled)	Several ANN regressors, each using current link utilization to determine on/off threshold	Focus learning/predictions on configuration options with the most impact
[93]	Dynamic link control (enabled/disabled and direction)	Per-node decision tree classifier using link, buffer, and data statistics	Try feature engineering using domain knowledge, then remove unnecessary information
[94]	Dynamic power scaling for photonic interconnect	Linear regression using broad list of common NoC features and photonic-specific feature	Use general features to accommodate more diverse designs
[95]	Dynamic link/router DVFS with power and thermal constraints	Classification w/several local ANNs and global ANN using task demands and constraints	Experiment with hierarchical control methods for sufficiently complex tasks
[96, 97]	Dynamic link/router DVFS for optimal energy-efficiency	Various nonlinear regression models and deep Q-learning using buffer util. or energy efficiency	RL can be more effective than supervised learning in determining thresholds for control tasks
[98]	Dynamic NoC routing	Q-learning using delivery time estimates	Machine-learning models can handle broad ranges of workloads better
[99]	Dynamic NoC routing for changing topology	Q-learning using delivery time estimates	Machine-learning models can better adjust to non-standard operating environments
[100]	Dynamic NoC routing for bufferless fault-tolerance	Q-learning using number of hops to estimate delivery time	Machine-learning models can better adjust to unforeseen operating environments
[101]	Dynamic NoC routing for high-performance w/ congestion	Q-learning using delivery time estimates to NoC region	Machine-learning models can enable greater flexibility in common operating environments

Table 3.4: (*Continued*) Summary of works on machine learning for NoCs (*Continues*)

[102]	Dynamic injection throttling for bufferless, multi-hop NoC	Q-learning using per-node injection queue occupancy	Some applications may require dedicated methods for global data collection
[103]	Dynamic injection throttling for hotspot prevention	Per-region ANN classifier using average buffer utilization	Supervised learning control can benefit from holistic training (learn impact of own actions)
[104]	Dynamic injection throttling for optimal performance	Centralized ANN classifier using several global packet statistics	Reduce overhead by combining control decisions when possible
[105]	Dynamic traffic arbitration	Deep Q-learning using wide range of features (not practical to implement)	Even impractical models can provide insight into margin for design improvements
[106]	" "	Train impractical deep Q-learning model, then extract insights for practical design	Some tasks benefit most from machine-learning insights, rather than the models themselves
[107, 108, 109]	Develop framework for 3D NoC design	Hill-climbing augmented by various regression models using current topology info	Multi-objective design can greatly benefit from machine learning predictive capabilities
[17]	Develop general framework for NoC design (and apply to routerless NoCs)	DRL using calculated hop count, supported by Monte-Carlo tree search	Modern RL models can explore vast design spaces exceeding 10^{100}
[111]	Develop general framework for NoC/SoC design space	Several regression models using wide range of design statistics and goals	Even high-level/abstract concepts in SoC design can be predicted w/machine-learning models
[113]	Improve NoC simulation tools	Support vector regression model using arrival rates, forwarding probability and waiting times	Combining traditional and machine learning-models can improve accuracy in some tasks

Table 3.4: (*Continued*) Summary of works on machine learning for NoCs

[114]	NoC and memory performance prediction	Cubic spline regression model using subset of design space	Recognize model bias introduced by random sampling in non-uniform architecture design spaces
[18]	Predict NoC link faults	Decision tree classifier using temperature, utilization, and wear-out	Modest model overhead can be outweighed by more efficient system operation
[115, 116]	Develop framework for NoC error mitigation, power-gating, and buffer control	Per-router Q-learning model using several router/buffer metrics and temperature	Machine-learning models can effectively coordinate diverse control options

3.5   SYSTEM-LEVEL OPTIMIZATION

3.5.1   ENERGY EFFICIENCY

Strict thermal and power limitations in modern computing systems have motivated continued research into energy efficiency optimization. In particular, machine-learning-based control schemes have shown promise in optimizing energy efficiency with minimal performance reduction, often enabling 60–80% reductions in the energy-delay product compared to race-to-idle schemes.

Bailey et al. [117] targeted power efficiency in heterogenous systems. Similar to Wu et al. [46], they clustered kernels by their scaling behavior to train multiple linear regression models. Runtime prediction uses two sample configurations, one from CPU execution and one from GPU execution, to determine the optimal configuration. Pan et al. [118] implemented a power management scheme using a multi-level RL algorithm. Their method propagates individual core states up a tree structure while aggregating Q-learning representations at each level. Global allocation is made at the root, then decisions are propagated back down the tree, enabling efficient per-core control. Won et al. [119] introduced a hybrid ANN + PI (proportional-integral) controller scheme for uncore DVFS. They initially trained the ANN offline, then refined predictions online using the PI controller. This hybrid scheme was shown to reduce the energy-delay product by 27% compared to a PI controller alone, with less than 3% performance degradation compared to the highest V/F level.

More recent work by Bai et al. [120] implemented an RL-based DVFS control policy adapted to a novel voltage regulator hierarchy using off-chip switching regulators and on-chip linear regulators. Individual RL agents adapt to a dynamically allocated power budget determined by a heuristic bidding approach. The design was enhanced using adaptive Kanerva coding [121] (to limit area/power overhead) and experience sharing (to accelerate learning). Chen and Marculescu [122] (later Chen et al. [123]) explored an alternative two-level strategy for RL-based DVFS. Similar to Bai et al. [120], they used RL agents at a fine-grained core level to select a voltage/frequency level based on an allocated share of the global power budget. They achieved further improvement by allocating power budget using a performance-aware, albeit still heuristic-based, variant that considers relative application performance requirements. Imes et al. [124] explored single-application system energy optimization for a broader range of configurations options including socket allocation, HyperThread usage, and processor DVFS. They identified several useful models, while noting that further work could optimize models and parameters. Analysis also provided insight into the benefit from single-model multi-resource optimization, particularly for neural networks. Finally, recent work by Tarsa et al. [125] considered a machine learning framework for post-silicon CPU adaptations using firmware updates to microcontroller-implemented models. Significant accommodations for statistical blindspots limit the rate of service-level-agreement violations while optimizing performance per watt for both general-purpose and application-specific deployment.

Several works have addressed energy-efficiency optimization for real-time workloads. Lo et al. [126] used linear regression to model execution time based on annotations and code features, enabling stricter service level guarantees at the cost of applicability when source code is unavailable. Mishra et al. [28] instead applied a comparatively complex hierarchical Bayesian model to combine both offline and online learning. In this approach, they accepted a high execution time penalty (0.8 s) in order to provide significantly more accurate predictions than online or offline training alone. This approach therefore targeted longer executing workloads, but can provide more than 24% energy savings over the next best approach. Later work by Mishra et al. [22] combined control theory and several machine-learning-based models. Their framework was realized by offloading learning to a server, allowing low overhead DVFS that reduces energy consumption by 13% compared to the best prior approach.

3.5.2   TASK ALLOCATION AND RESOURCE MANAGEMENT

In addition to energy control, machine learning offers an approach to allocate tasks to resources or resources to tasks by predicting the impact of various configurations on long-term performance.

Lu et al. [127] proposed a thermal-aware Q-learning method for many-core task allocation. The agent considered only current temperature (i.e., no application profiling or hardware counters), receiving higher rewards for task assignments resulting in greater thermal headroom. Evaluation indicated an average 4.3°C reduction in peak temperature compared to a heuristic approach. Work by Zheng et al. [128] explored multi-level scheduling in heterogeneous processing clusters. They proposed a DRL framework to divide video workloads, first selecting a worker node and then selecting a processing unit (CPU/GPU). The two DRL models act separately, but still work together to optimize overall throughput.

Allocating resources to tasks is another possible approach. Early work by Bitirgen et al. [129] considered a system with four cores and four concurrent applications. In their approach, per-application ANN ensembles predicted IPC for 2,000 configurations at each interval (500 K cycles). IPC predictions were then aggregated to choose the highest performing overall system configuration. This ensemble-based approach enables estimates for prediction confidence, but introduces scaling concerns, particularly with exponentially increasing configuration spaces. Jain et al. [130] explored an alternative approach with RL-based optimization of core DVFS, uncore DVFS, and dynamic LLC partitioning in a traditional multi-core processor. They used individual agents for each resource, potentially limiting co-optimization opportunities, while reconfiguring at relatively large 1B instruction intervals. Evaluation nevertheless indicated noteworthy reductions in energy-delay-product through multi-resource optimization.

Additional work has continued to consider low-level co-optimization in more diverse architectures. In particular, Nemirovsky et al. [131] introduced a method for IPC prediction and task scheduling on a heterogeneous architecture. They predicted IPC for all task arrangements using ANNs, then selected the arrangement with the highest IPC. Evaluation highlighted significant throughput gains ($> 1.3\times$) using a deep (but high overhead) neural network, indicating one possible application for pruning (discussed in Section 6.2). Ma et al. [132] considered optimization for an energy-harvesting nonvolatile processor (NVP) involving ten hardware resources as well as frequency. They trained a separate neural network for each resource, thus potentially limiting co-optimization opportunities (similar to [130]), but still observed forward progress[3] improvements of roughly $2\times$ on average.

Several related works specifically target resource allocation under strict quality of service (QoS) guarantees when co-scheduling high-priority, latency-critical workloads with low-priority, best-effort workloads. Nishtala et al. [133] proposed an RL scheme to control core allocation and core frequency in a heterogeneous system. The RL model is trained online using a simple state machine, first learning to minimize QoS violations and then to maximize performance or minimize power. Later work [134] again considered core allocation and core frequency, but with a deep Q-learning approach. In this approach, they integrated hardware performance counters to better determine when QoS may be violated, thus enabling significant power reductions with relatively limited impact to QoS.

Finally, work by Ding et al. [135] established a somewhat contradictory trend between model accuracy and system optimization goals, showing that improvements in accuracy are not necessarily beneficial. They additionally propose a method to build better models by accounting for the problem structure (e.g., focus sampling on the optimal front), resulting in lower accuracy but higher real-world energy savings.

---

[3]Forward progress in NVPs considers both executed instructions and the number of backups (due to power loss) that hinder progress.

Table 3.5: Summary of works on machine learning for system-level optimization (*Continues*)

Work	Goal	Strategy	Lessons Learned
[117]	Optimize performance in power-constrained system	Linear regression on CPU/GPU configuration options	Explore clustering on workload behaviors to simplify predictions
[118]	Optimize energy efficiency w/minimal performance reduction	Multi-level Q-learning using current core state to determine next core state	Experiment with hierarchical control methods for sufficiently complex tasks
[119]	" "	Combined ANN classifier and control-theoretic controller using NoC latency metric	Combined control schemes can minimize detrimental aspects of machine-learning models
[120]	" "	Q-learning on instruction mix and cache/memory statistics to decide frequency	Consider experience-sharing methods to speedup online learning
[122, 123]	" "	Per-core Q-learning agent using IPC, MPKI, and current power/frequency	Consider multi-objective reward structures for more general improvements
[124]	" "	Multiple classification models using common hardware performance counters	Multi-parameter predictions can benefit from unified model in modest exploration spaces
[125]	Optimize energy efficiency w/minimal service-level-agreement violations	Multiple classification models on common performance counters	Test as many workloads as possible to ensure reliable performance
[126]	Optimize energy efficiency for real-time workloads	Linear regression on code features to predict instruction count	Consider trade-offs between prediction time (allowing faster control) and accuracy
[28]	" "	Hierarchical Bayesian model using power and performance on sampled configurations	Modest model overhead can be outweighed by more efficient system operation
[22]	" "	Various regression models using power and latency on sampled configurations	Combined traditional and machine-learning models can improve performance and ease implementation

Table 3.5: (*Continued*) Summary of works on machine learning for system-level optimization

[127]	Optimize thermal headroom under varying workload intensities	Q-learning using current temperature readings to determine task assignment	Machine learning models tolerate approximate implementations, thereby reducing overhead
[128]	Optimize task throughput in hybrid CPU-GPU clusters	Two-level DRL using task characteristics and current resource state	Transfer learning can greatly improve practical application of (originally) task-specific models
[129]	Optimize multi-app performance	Per-app ANN ensemble regression on hardware performance counters and resource allocation	Ensemble methods can be made practical by multiplexing calculations
[130]	Optimize multi-app energy efficiency	Per-resource Q-learning agents using time per instruction to determine next allocation	Consider experience-sharing methods to speedup online learning
[131]	Optimize multi-app performance on heterogeneous architecture	Per-core-type (big/little) ANN regression on common perf. counters	Consider separate models for sufficiently distinct perf. prediction scenarios
[132]	Optimize performance on energy-harvesting processor	Per-resource ANN regression on current resource utilization	Modest model overhead can be outweighed by more efficient system operation
[133]	Optimize energy-efficiency for coscheduled latency-critical apps	Q-learning using QoS statistics and common hardware performance counters, augmented by traditional control	Combined traditional and machine-learning models can improve performance and ease implementation
[134]	" "	Deep Q-learning using common hardware performance counters to adjust resources	Consider multi-headed models to limit model complexity with multi-dimensional predictions
[135]	Examine learned behaviors in model for energy-efficiency optimization	Hierarchical Bayesian model using subset of configurations to predict performance/power	Higher model accuracy does not necessarily coincide with better real-world application
[16]	Explore hardware-based malware detection	Logistic and ANN regression on instruction mix (opcodes)	Explore multiple feature selection methods to better understand importance

3.5.3   SECURITY

Malware detection, a traditionally software-based task, has been explored using machine learning to enable reliable hardware-based in-execution detection. For example, Ozsoy et al. [16] tested both logistic regression (LR) and neural network classifiers trained on low-level hardware counters. Optimizations based on task-specific feature selection and reduced model precision result in high accuracy (100% malware detection and less than 16% false positives) with minimal overhead (0.04% core power and 0.19% core logic area) for the LR model. These promising results suggest future opportunities to integrate machine learning into computer security.

3.6   APPROXIMATE COMPUTING

Approximate computing has many facets, such as circuit level approximations (e.g., reduced precision adders), control level approximations (e.g., relaxed timings), and data level approximation. Methods using machine learning generally fall within the last category, offering a powerful function/loop approximation technique that commonly provides 2–3 times application speedup and energy reduction with limited impact on output quality.

Esmaeilzadeh et al. [136] introduced a neural processing unit (NPU), which provides a novel approach for programmable approximation using neural networks. They developed a framework to realize Parrot transformations that translate annotated code segments into neural networks approximations. Tightly integrating the NPU with the CPU allows an average 2.3× speedup and 3.0× energy reduction in studied applications. This framework was later extended by Yazdanbakhsh et al. [137] to implement neural approximation on GPUs. Neural approximation was integrated into the existing GPU pipeline, enabling component re-use and approximately 2.5× speedup and 2.5× reduced energy. Grigorian et al. [138] presented a different approach for a multi-stage neural accelerator. Inputs are first sent through a relatively low accuracy/overhead neural accelerator, then checked for quality; acceptable results are committed, while low quality approximations are forwarded to an additional, more precise, approximation stage. The problem with these works is that error is either constant [136, 137] or requires several stages with potentially redundant approximation [138]. For that reason, Mahajan et al. [139] introduced MITHRA, a co-designed hardware-software control framework for neural approximation. MITHRA implements configurable output quality loss with statistical guarantees. Specifically, machine learning classifiers predict individual approximation error, allowing comparison to a quality threshold. Recent work by Oliveira et al. [140] also explored approximation using low-overhead classification trees. Even with software-based execution, they achieved application speedup comparable to an NPU [136] hardware implementation. Finally, machine learning has also been used to mitigate the impact of faults in existing approximate accelerators. Taher et al. [141] observed that faults tend to manifest in a similar manner across many input test vectors. This observation enables effective error compensation using a classification/regression model to correct output based on predicted faults for a given input.

Table 3.6: Summary of works on machine learning for approximate computing

Work	Goal	Strategy	Lessons Learned
[136, 137]	Predict code segment output w/machine-learning model	ANN regression on subset of inputs	Explore balance between dedicated machine-learning hardware and re-use of existing hardware
[138]	Predict code segment output w/machine-learning model	Multi-level ANN regression on subset of inputs	Consider multiple models to achieve different performance/accuracy levels
[139]	Provide statistical guarantees for machine-learning function approximations	ANN and decision tree classifiers using input values to predict acceptable error	Need to consider impact of mispredictions in diverse tasks, not just function approximation
[140]	Predict code segment output w/machine-learning model	Decision tree classifier using representative input/output values	Classication-based value prediction can greatly improve efficiency with minor accuracy penalty
[141]	Mitigate impact of errors in accelerators	Decision tree classifier using representative inputs to predict common fault corrector	Modest model overhead can be outweighed by more efficient system operation

SUMMARY

In this chapter, we have expanded our perspective on the diverse range of possibilities for AI application in architecture. Specifically, we have observed how machine learning can be applied in system simulations to reduce execution time by efficiently approximating system and application behaviors. Similarly, these models can enable new possibilities in tasks such as cross-architecture performance and power prediction. These capabilities, paired with the broad applicability of machine learning algorithms/models, have produced state-of-the-art advancements in practically all major architectural components, including the core, cache/memory, the NoC, and GPUs. In particular, we have seen how machine learning models can enable efficient design space exploration, perform task or resource scheduling, proactively control both low- and high-level architectural resources, predict data use, optimize system energy-efficiency, and much more. These machine-learning-based approaches are usually highly generalizable and can accommodate changes in application behavior or operating demands, thus providing distinct advantages over traditional design strategies. Emerging applications, such as machine-learning-enabled approximate computing, provide further indication of the promising future for AI application to architecture. In our view, the existing works covered in this chapter represent a small fraction of possible applications that may emerge from this rapidly developing field in the near future. Continued advancements in both machine learning and architecture may lead to many other exciting new possibilities.

Practical implementations for the applications discussed in this chapter require careful consideration, both when selecting/building a machine learning model and when integrating any model into real-world architecture designs. In the next chapter, we continue to build intuition regarding these practical considerations as we examine three case studies that represent the state-of-the-art in particular aspects of machine learning applied to computer architecture.

CHAPTER 4

Case Studies

The previous chapter highlighted the broad range of machine learning applications in architecture and hinted at a number of considerations for specific tasks, implementation requirements, etc. In this chapter, we work to expand these principles as we scrutinize three exemplars of machine learning application in architecture. Each case study covers a different learning approach and architectural sub-system. These case studies therefore serve as an opportunity to observe the real-world results and ramifications of noteworthy design strategies and implementation choices.

The first case study examines supervised learning application in branch prediction, representing one of a few mainstream applications in current practice. The second case study explores reinforcement learning application to vast design space exploration problems, with a focus on NoCs. Finally, the third case study investigates unsupervised learning application in memory systems and a noteworthy hardware implementation.

4.1 SUPERVISED LEARNING IN BRANCH PREDICTION

One of the earliest applications of AI to computer architecture is dynamic branch prediction. Microprocessors mitigate control hazards by using branch predictors to speculatively fetch and execute instructions beyond conditional branches. The penalty of a mispredicted branch is proportional to the size of the instruction window and often exceeds the penalty of a first- or second-level cache miss. Thus, branch predictor accuracy is very important for good performance.

Prior to the application of AI to the branch prediction problem, most branch predictors were based on refinements to two-level adaptive branch prediction [142]. This scheme is based on branch history: a shift-register keeps the outcomes of a fixed number of recent conditional branches, recording 1 for taken branches and 0 for not-taken branches. This first level of history indexes a second level of saturating counters. The high bit of the counter gives the prediction, 1 for taken and 0 for not-taken. When the branch outcome becomes known, the corresponding counter is incremented if the branch were taken, decremented otherwise. Thus, the correlation between branch history and branch outcome is used to predict branches.

4.1.1 BRANCH PREDICTION WITH PERCEPTRONS

In 2001, Jiménez and Lin proposed using simple perceptrons to replace the second level of saturating counters. As of this writing, perceptron-based branch predictors are used in high-

function prediction (pc: **integer**): { taken, not_taken };
begin
    // Hash the pc to select a row of W
    i = pc mod n
    // Compute output of ith perceptron using G[1..h] as input

$$y_{out} = W[i, 0] + \sum_{j=1}^{h} \begin{cases} W[i,j] & \text{if } G[j] = \text{taken} \\ -W[i,j] & \text{if } G[j] = \text{not\_taken} \end{cases}$$

    // Make the prediction based on the sign of y_{out}
    if $y_{out} \geq 0$ **then**
        prediction := taken
    else
        prediction := not_taken
    end if
end

procedure train (i, y_{out}: **integer**; prediction, outcome: { taken, not_taken });
    // If incorrect or y_{out} below threshold then adjust weights
    if prediction $\neq$ outcome or $|y_{out}| \leq \theta$ **then**
        // Increment bias weight if taken, decrement if not_taken

$$W[i, 0] := W[i, 0] + \begin{cases} 1 & \text{if outcome} = \text{taken} \\ -1 & \text{if outcome} = \text{not\_taken} \end{cases}$$

        for j **in** 1...h **in parallel do**
            // Increment j^{th} weight for positive correlation, decrement for negative correlation

$$W[i, j] := W[i, j] + \begin{cases} 1 & \text{if outcome} = G[j] \\ -1 & \text{if outcome} \neq G[j] \end{cases}$$

        end for
    end if
    // Update the global history shift register
    G := (G << 1) **or** outcome
end

Figure 4.1: Perceptron prediction and update algorithm.

performance ARM processors from Samsung [143, 144] as well as processors from AMD and Oracle [145, 146].

    Jiménez and Lin's perceptron branch predictor uses perceptron learning [147, 148] to predict the directions of conditional branches [14, 149]. We review the design of the perceptron predictor, describing algorithms using an Algol-like pseudocode with keywords in **boldface** and comments in *italics*. We use *taken* and *not_taken* as meaningful names for Boolean constants.

    The perceptron predictor is similar to other predictors in that it keeps a global history shift register that records the outcomes of branches as they are executed, or speculatively as they are predicted. The width of this register is the history length for the predictor, hereafter referred to as h.

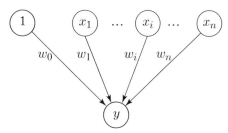

Figure 4.2: Perceptron model. Input values $x_1, \ldots, x_n$, are propagated through weighted connections by taking their respective products with weights $w_1, \ldots, w_n$. These products are summed, along with the bias weight w_0, to produce the output value y.

4.1.2   PERCEPTRON PREDICTION EXAMPLE

Figure 4.2 shows a graphical model of a perceptron. A perceptron is represented by a vector whose elements are the weights. For our purposes, the weights are signed integers. The output is the dot product of the weights vector, $w_{0...n}$, and the input vector, $x_{1...n}$ (x_0 is always set to 1, providing a "bias" input). The output y of a perceptron is computed as

$$y = w_0 + \sum_{i=1}^{n} x_i w_i. \tag{4.1}$$

The inputs to the perceptron, $x_1 \ldots x_n$, are from the global branch history, with x_i representing the outcome of the ith most recent branch. The inputs are *bipolar*, meaning that each x_i is either -1, indicating *not taken* or 1, indicating *taken*. A negative output is interpreted as *predict not taken*. A non-negative output is interpreted as *predict taken*.

Figure 4.3 illustrates the concept of a perceptron producing a prediction and being trained. The perceptron weights vector is multiplied by the branch history, and summed with the bias weight, 1 in this case, to form the perceptron output. In this example, the perceptron incorrectly predicts that the branch is taken. Once the outcome of the branch becomes known, training adjusts the weights when it discovers the misprediction. With the adjusted weights, assuming that the history is the same the next time this branch is predicted, the perceptron output is negative, so the branch will be predicted not taken.

4.1.3   ORGANIZATION OF THE PERCEPTRON PREDICTOR

The perceptron predictor keeps an $n \times (h + 1)$ matrix $W[0 \ldots n - 1, 0 \ldots h]$ of integer weights, where n is a design parameter and each row is a single perceptron as described above. Weights are typically 8-bits. Each row of the matrix is an $(h + 1)$-length *weights vector*. Each weights vector stores the weights of one perceptron that is controlled by perceptron learning. In a weights vector $w[0 \ldots h]$, the first weight, $w[0]$, is the bias weight. Thus, the first column of W contains the bias

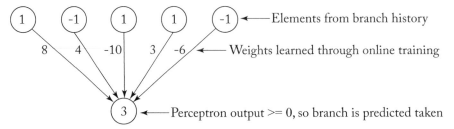

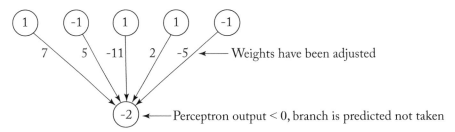

Figure 4.3: Perceptron prediction and training.

weights of each weights vector. The Boolean vector $G[1 \ldots h] \in \{1 \ldots h\} \times \{taken, not\_taken\}$ represents the global history shift register.

Prediction and Update Algorithms

Figure 4.1 gives pseudocode for the prediction and update algorithms for the original perceptron predictor. The *prediction* algorithm returns a Boolean value predicting the branch at address *pc*.

When a branch outcome becomes known, the *train* algorithm is invoked to update the predictor. The training algorithm takes an integer parameter θ that controls the trade-off between long-term accuracy and the ability to adapt to phase behavior. It has been empirically determined that choosing $\theta = \lfloor 1.93h + 14 \rfloor$ gives the best accuracy [149]. Thus, θ is a constant for a given history length. Once the outcome of a branch becomes known, the algorithm is used to update the perceptron predictor, taking as parameters the outcome as well as the values of i, *prediction*, and y_{out} computed during the prediction phase.

Thus, the bias weight is incremented (decremented) if the branch is taken (not taken), while the rest of the weights are incremented (decremented) if the branch outcome is equal (not equal) to the corresponding bit of the global history shift register.

Implementation

We review some of the suggestions for a practical implementation of the perceptron predictor.

The matrix W should be implemented as a tagless direct-mapped memory of n blocks with the ith block containing h 8-bit weights that form the weights vector of the ith perceptron. Thus, each time a prediction is needed, the weights vector corresponding to that value of pc is read from memory.

The computation of y_{out} can be arranged as a Wallace-tree [150] adder to add the summands. This allows the circuit performing this computation to have a depth of $O(\log h)$ gate delays, as opposed to $O(h)$ gate delays with a naive summing algorithm.

4.1.4 IMPROVING THE PERCEPTRON PREDICTOR

One disadvantage of the original perceptron predictor was its high latency. Even using the high-speed arithmetic tricks mentioned above, the latency of the computation of y_{out} is high relative to the clock period of a deeply pipelined microarchitecture. It has been shown that performance is highly sensitive to high branch predictor latency [151], even when special techniques are used to mitigate latency [152].

Another disadvantage is that perceptrons are unable to learn functions that are not linearly separable, i.e., if the space of histories for *taken* vs. *not-taken* cannot be separated by a linear decision surface, the perceptron predictor will not be able to accurately predict the branch. In practice, the perceptron predictor has good accuracy, but the accuracy can be greatly improved with some simple techniques to reorganize the predictor.

Jiménez proposed using the path of branch addresses leading to a branch to select weights used for the dot-product computation, rather than using the same set of weights each time a given branch is encountered [153, 154]. This path-based approach improves accuracy and gets around the linear separability problem by allowing the path to allocate as many or as few perceptron weights as needed to represent the different control-flow contexts leading to a branch. It also greatly relieves the latency problem by allowing a pipelined implementation of the dot-product: the summation of weights may begin as soon as the least recent bit in the history is known, rather than all at once when the prediction is needed.

4.1.5 CURRENT APPROACH

The current approach to perceptron-based branch prediction in modern processors is known as the *hashed perceptron* [155]. The idea is to have several tables, each indexed by a different hash of branch history. The tables have somewhat wide saturating confidence weights. The selected weights are summed and the prediction is taken if the sum is at least zero, not-taken otherwise. On a mispredict or low-confidence correct prediction, the corresponding weights are incremented if the branch is taken, decremented otherwise. The hashed perceptron predictor is similar to an idea of Loh and Jiménez called *modulo path-history* [156], while O-GEHL [157] is a very specific instance of the general technique. The hashed perceptron predictor, like modulo path-history and O-GEHL, improves over the original perceptron predictor [14] by breaking the one-to-one correspondence between weights and history bits, allowing a more efficient

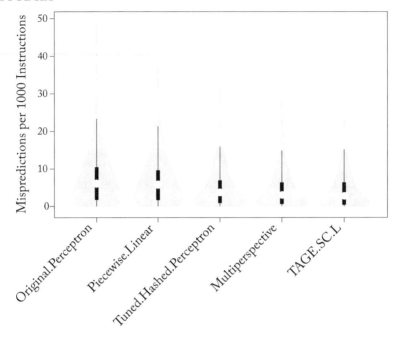

Figure 4.4: Accuracy of perceptron-based predictors.

representation as well as a less complex adder tree implementation. The O-GEHL approach of using geometric history lengths led to the TAGE family of branch predictors [21]. TAGE is not a perceptron-based predictor, but rather uses prediction by partial matching in a set of tagged tables. However, subsequent iterations of TAGE returned to using a perceptron predictor styled as a "statistical corrector" indexed by a variety of features to correct unconfident TAGE predictions [19].

The violin plot in Figure 4.4 illustrates the accuracy of several branch predictors across 200 traces, which were randomly chosen from the 5th Championship Branch Prediction competition [158]. Here, the y-axis gives the number of mispredictions per 1000 instructions (MPKI) and the light grey area represents the probability density (i.e., the fraction of traces that exhibit a certain MPKI). Branch predictors on the x-axis include the original perceptron predictor, a generalization of perceptron prediction called piecewise linear branch prediction [154], a tuned hashed perceptron predictor similar to an industrial design [159], a multiperspective perceptron predictor [20], and TAGE-SC-L. Moving from left to right, we see that prediction accuracy has continued to increase over the years. In particular, we also observe that variance in prediction accuracy has decreased along with MPKI; the bottom-heavy shape of the plots for later predictors indicates that most benchmarks are easily predicted by recent perceptron-based predictors.

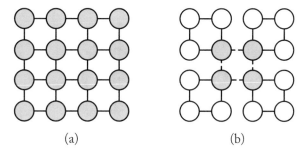

Figure 4.5: Router-based NoC architectures: (a) mesh and (b) hierarchical-ring. Nodes (circles) with significant routing hardware are colored in blue.

4.2   REINFORCEMENT LEARNING IN NOCS

Most RL applications in architecture involve real-time, online control tasks with modest de-sign spaces. These works highlight the flexibility of RL, yet do not realize the full extent of reinforcement learning design space exploration capabilities. Recent work by Lin et al. [17] in-stead applies RL at design-time to explore the vast architectural design space of NoCs. Using a framework of recent machine learning innovations, the work showcases how reinforcement learning can successfully explore architectural design spaces exceeding 10^{100} for current NoC sizes. We present this work here to provide insight into a new area with strong potential for further applications.

4.2.1   BACKGROUND

NoCs have become the backbone for communication in ever-expanding many-core processors. As such, researchers have proposed diverse designs that enable broad trade-offs based on per-formance characteristics, such as latency, throughput, and hop count, as well as practical con-straints, such as wiring, power, area, and overall manufacturability. We briefly detail fundamental NoC designs and recent routerless NoCs, which are the foundation for this case study.

Router-Based NoCs
Router-based NoCs have long been a typical design option for a variety of reasons. The mesh NoC, illustrated in Figure 4.5a, is a notable example that has become a *de facto* standard, in part, due to its extremely regular/scalable design. In this mesh design, all nodes are identical, with each node containing both a processing element and a router. Data can therefore be sent, received, or forwarded at each node. Mesh NoCs also provide high reliability when combined with adaptive routing algorithms since there are many possible paths between any two nodes. Unfortunately, these benefits come with a cost. Depending on performance requirements, mesh

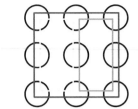

Figure 4.6: Routerless NoC architecture. Each rectangle represents one loop along which data can be forwarded.

NoCs can incur power and area overhead that represents a significant portion of the total chip resources (11% area [160] and up to 28% power [161, 162]).

Hierarchical-ring NoCs, illustrated in Figure 4.5b, are another router-based option. This design instead features a hierarchy of local nodes connected in a ring and a global ring that connects these four sets. Observe that routers (shown in blue) are only required at nodes intersected by the global ring since communication within a local group is possible through simple node-to-node forwarding. All other nodes only require buffers and control logic to inject/eject data from the ring. Evidently, router overhead is lessened, yet this scheme introduces potential bottlenecks at connecting points with the global ring; assuming unidirectional rings, there is only a single path from the bottom-left node to the top-right node.

Routerless NoCs

Routerless NoCs [163] were recently proposed to address the limitations and reduce overhead associated with traditional router-based NoCs. Routerless NoCs, like hierarchical-ring designs, rely upon a collection of loops to enable communication. Routerless NoCs differ from hierarchical-ring, however, in that loops can span the entire NoC and many multi-loop combinations are possible. One possible realization for a 3×3 routerless NoC is shown in Figure 4.6. Together, these three loops connect every node to every other node. Therefore, using this architecture, each node has a direct forwarding path to any other node so there is no need for routing hardware connecting individual loops. The only "routing" is loop selection, which can be performed entirely at the source node using a small lookup table. As a result, these routerless NoCs achieve substantial power and area savings while also improving performance compared to mesh NoCs. In fact, the routerless NoC designs generated by the framework in this case study achieve a 3.25× increase in throughput, 1.6× reduction in packet latency, 5× reduction in power, and 82% reduced area compared to a typical 8×8 mesh [17].

Routerless NoCs offer scaling performance by increasing either the number of wires in a loop or the total number of loops, thus offering additional design flexibility. The main constraint is that wires are, ultimately, a finite resource that must be limited for practical chip layout. Routerless NoC design therefore mandates efficient loop placement: an "optimal" routerless

NoC design should contain a finite set of loops that achieves the lowest average hop count out of all possible loop combinations while adhering to a wiring constraint. In the paper by Lin et al. [17], on which this case study is based, the authors advocate a constraint on node overlapping (i.e., the number of loops "passing through" a node). Two prior papers [163, 164] proposed general design strategies for routerless NoCs, but neither provided a reliable method to explore loop configurations and design routerless NoCs under various design constraints.

4.2.2 MOTIVATION FOR MACHINE LEARNING

The primary motivation for machine learning in routerless NoC design (and NoC design as a whole) is the immense design space for possible loop configurations, coupled with the need for strict constraints. For example, the design space for 8×8 routerless NoCs includes $\binom{784}{50} \approx 10^{79}$ designs when selecting 50 loops from 784 possible rectangular loops [17]. Increasing NoC size in many-core processors pushes this design space even further, approaching $\binom{23409}{150} \approx 10^{392}$ for an 18×18 routerless NoC. Some of these designs will violate connectivity requirements (i.e., not connect every node to every other node) while others will violate wiring constraints. Nevertheless, the design space remains many orders of magnitude beyond the capabilities of an exhaustive search, especially as NoC sizes continue to scale.

Traditional search methods can be applied to this task, but past works have shown that there are still fundamental limitations when exploring this vast a design space. One prior work [164] adopted an evolutionary approach using a genetic algorithm to search for high-performing routerless NoC loop configurations. Each generation of individuals (i.e., current loop configurations) is selected using an objective function, but evolution relies on random mutation, resulting in unreliable and relatively slow searches [163] as well as undesirable design characteristics, such as high average hop count and long loops. Design constraints could be built into the fitness, but the search may violate these constraints to achieve higher performing NoCs. There may also be issues with this genetic approach being unable to find viable solutions as the NoC size increases.

Additional related work on 3D NoC design [109] (mentioned in Chapter 3) provides further insight by comparing their machine-learning-based method against several other traditional search schemes including branch-and-bound and simulated annealing. The first algorithm, PCBB [165], is an improved branch-and-bound method for multi-objective NoC design. They found that PCBB required over two orders of magnitude more time to achieve the same NoC design quality as their proposed STAGE algorithm (a supervised-learning-enhanced hillclimbing algorithm). The second search algorithm, AMOSA [112], is a multi-objective search algorithm based on simulated annealing. AMOSA found high-quality NoC designs much faster than PCBB, but still required approximately an order of magnitude more time than the proposed algorithm when optimizing four design criteria. This difference in search time between AMOSA and the ML-based STAGE algorithm increases exponentially as the design space/complexity grows, thus indicating how machine learning can improve performance compared to more traditional algorithms.

The choice between various machine learning algorithms for design space exploration and general NoC design is somewhat less well-defined. Nevertheless, some guidance can be found in the recent success of AlphaGo Zero [3] in complex games. The DRL algorithms used by AlphaGo Zero successfully navigated a design space of approximately 10^{170}, learning optimal actions by searching a Monte Carlo tree that incorporates actions from previous games. This DRL approach effectively combines search and prediction, thereby providing a promising approach.

4.2.3 IMPLEMENTATION

Limitations in traditional search options relating to performance and reliability motivate a new approach for architectural design space exploration. To that end, we now discuss the DRL framework [17] and its application to routerless NoC design. We begin by introducing the overall framework, then highlight three significant considerations: the state/action/reward structure, the neural network architecture for the learner, and the exploration approach featuring a Monte Carlo tree search. As we will see, each component plays a key role in enabling efficient and reliable scaling with the NoC size.

Framework Overview

The DRL design process, depicted in Figure 4.7, does not require any *a priori* information. Instead, in this setup, the DNN and Monte Carlo Tree Search (MCTS) gradually learn on their own how to generate high-performing routerless NoC designs by exploring individual loop placements. As mentioned earlier, each complete routerless NoC design requires many individual loop placements (indicated by the loop going from "DNN" to "MCTS" and back to "DNN" when "More Actions" is selected). Similarly, the framework may need to explore many complete routerless NoC designs in order to learn optimal design strategies (indicated by the loop back to "Blank Design").

Framework execution begins with a single blank design. An initial loop placement is suggested by the DNN and then MCTS is used to quickly locate promising branches (i.e., unexplored routerless NoC configurations). Assuming more valid actions are possible, the framework loops back to the DNN, which selects another action based on current state information. At each step, the MCTS acts as a guide to ensure that the search maintains a balance between exploring new actions and exploiting known high-performing actions. Eventually, there will be no more possible valid actions due to design constraints, at which point the DNN and search tree are updated. Observe that these updates are based on an intermediate, simplified metric rather than execution time in full-system simulations. Even though the overall goal is reduced execution time, it is not feasible to simulate every generated routerless NoC design since every full-system simulation may take several hours. Hop count is instead used here as the intermediate metric

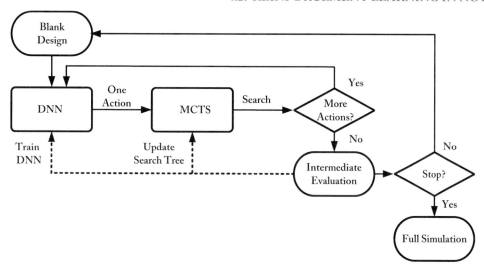

Figure 4.7: Deep reinforcement learning framework.

since it is closely correlated to execution time, yet can be calculated many orders of times faster. This overall process repeats many times as diverse routerless NoC configurations are explored. Finally, once a satisfactory design is generated, full-system simulations can be used to verify real-world performance. This design procedure is both reliable and easily generalizable, thus this framework could be broadly applied to problems beyond routerless NoC design.

Reinforcement Learning Structure (State/Action/Reward)

The DRL approach chosen for this work involves a DNN, thus mandating a fixed-sized representation for states and actions that is consistent throughout the design process. This requirement is challenging as the current loop configuration cannot be directly used for the state representation because the loop configuration is continuously changing while the NoC is being constructed. Instead, an alternative representation can be based on current node-to-node hop count, which is derived from the current loop configuration. Hop count can be calculated for nodes that are connected at the current design step. Meanwhile, nodes that are not connected at the current design step can default to a much larger "unconnected" hop count value (e.g., $5N$ as discussed shortly).

Representation for actions must, likewise, be a fixed-size to accommodate a DNN-based implementation for the learner. Consequently, we cannot simply list the nodes in a loop since loop lengths may vary. One possible solution (not used in the paper) involves marking all nodes across the NoC as either connected or unconnected by each new loop. This representation would allow non-rectangular loops, although an additional search step would still be required since there is not a one-to-one mapping between the action representation and individual loops. The issue with this approach is that it leads to a highly complex action space that may not scale well with increasing NoC size. This paper instead assumes rectangular loops, resulting in a far simpler representation; each rectangular loop can be represented as just two points (x-y coordinates) that define the opposing corners of the loop. This representation also provides a one-to-one mapping between the two points and individual loops, so no further translation is required.

At first glance, the reward structure need not be directly bound by the environment (i.e., the NoC size). This is not, however, strictly true. In general, penalties are enforced for negative actions, with the goal of discouraging all non-productive actions (e.g., duplicate loops) or actions that violate design constraints [17]. Once the NoC is completed, a final penalty, scaling with hop count, is applied so that the agent is encouraged to produce fully-connected NoC with low average hop count. Observe that for an $N x N$ routerless NoC (i.e., a 2D NoC with N nodes in both dimensions), the largest rectangular loop passes through $4N - 4$ nodes, thus the worst-case hop count will be $4N - 3$. If the agent receives a final penalty less than $4N - 3$ for unconnected nodes, it may exclude the worst-case loop to minimize overall penalties. The penalty of $5N$ therefore ensures that all valid loops are favored over rule violations. A similar situation applies for the overlapping constraint penalty, since the agent may otherwise learn to place an additional loop around the outer edge to reduce the $4N - 3$ worst case hop count.

Neural Network Architecture

Another key element in the effectiveness of the proposed framework (as well as in AlphaGo Zero) is the two-headed neural network architecture, illustrated in Figure 4.8, that combines both the policy and value functions into a single network with two heads. The first N layers are functionally identical to a traditional neural network. Following that, the network splits into two heads, each of which takes the output from layer N and specializes in learning for a particular task. Formally, this is referred to as multi-task learning [166].

The intuition for multi-task learning is that related tasks likely share information necessary to make a prediction. In this particular case, it is not hard to imagine that the information required to make a high-quality action and predict the value of a given state will be correlated, thus shared layers will be encouraged to learn information that is relevant to both these tasks. Note that the backpropagation algorithm requires minimal change; starting from the output of each head, all layers in that head and all shared layers will be updated.

Exploration via Monte Carlo Tree Search

Including MCTS in the framework might initially seem counterproductive given the issues discussed with traditional search schemes. Nevertheless, both AlphaGo Zero [3] and the work in this case study demonstrate how MCTS and DRL can play a synergistic role in the overall design process. Specifically, the DNN generates coarse-grained designs that meet constraints while the MCTS helps to efficiently refine designs and optimize performance.

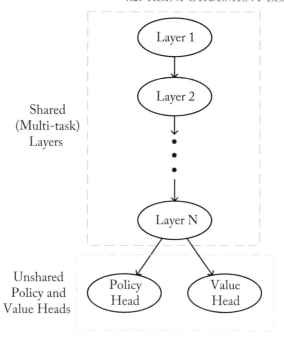

Figure 4.8: Two-headed neural network architecture.

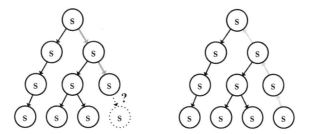

Figure 4.9: Monte Carlo tree search.

MCTS structure, illustrated in Figure 4.9, consists of nodes, representing routerless NoC loop configurations, and edges, representing individual loop additions. In other words, children of a specific parent node will have identical loop configurations (that of the parent) except for one additional unique loop. This organization naturally summarizes the benefits of particular loops by comparing all children of a specific parent.

MCTS involves three phases: search, expansion+evaluation, and backup.

Search: The search phase, shown on the left half of Figure 4.9, begins from the current loop configuration and rapidly selects additional loops based on past knowledge (PUCT algorithm [167]) and an epsilon-greedy selection scheme: the best known option is selected with probability $1 - \epsilon$ while a greedy action is selected with probability ϵ. In this work, the greedy action is defined as a loop addition that connects as many previously unconnected nodes as possible while adhering to design constraints. This action may select large loops early in a design, but helps guide design toward fully connected NoCs. Regardless, alternative definitions are possible. This search phase ends when a leaf node (i.e., a previously unexplored configuration) is reached.

Expansion+evaluation: In this phase, the current loop configuration is passed to the DNN and a new edge (i.e., loop) is selected, thus expanding the tree. Note that no rollout (i.e., random search until design completion) is performed after each DNN loop addition. Instead, we simply record the DNN action probability and the reward, both of which are used in the next step.

Backup: This final phase, illustrated on the right side of Figure 4.9, acts to update the search tree using information gained from the current search (e.g., loop rewards). The updated search tree can then more accurately guide later searches toward "optimal" actions. Implementation in this phase differs slightly for a single-threaded vs. multi-threaded approach. For the multi-threaded version of the framework [17], updates for many search/expansion+evaluation steps are combined and performed once a design is completed, allowing all these updates to reflect the final holistic hop count reward.

4.2.4   RESULTS

The proposed framework, referred to as "DRL," generates high-performance routerless NoC designs that offer a number of improvements over prior work. Specifically, for an 8×8 NoC, DRL reduces packet latency by 1.62× and increases throughput by 3.25× compared with a typical mesh design. Furthermore, these improvements are shown to scale well with NoC size; when scaling from 4×4 to 10×10 NoC size, throughput for a typical mesh design decreases by 59% while DRL throughput decreases by just 4.7%. The main reason for these improvements is that routerless NoCs inherently offer reduced per hop latency (one cycle) compared to typical mesh designs, which require two cycles or more. Prior routerless NoC designs could not, however, fully exploit these benefits since average hop count in those works is much higher than in mesh (1.5× in the prior state of the art [163]). DRL, on the other hand, has only slightly increased hop count compared with mesh, leading to higher overall performance.

Routerless NoC designs generated by DRL also provide substantial power and area savings. As mentioned previously, routerless NoC designs require just a small lookup table to perform routing when a packet is first injected into the network, whereas mesh requires several complex components (e.g., a crossbar and supporting arbitration logic) to perform routing at every hop. Consequently, DRL reduces total power consumption by 80% and area by 82% when compared with a typical mesh design.

The proposed framework offers additional utility through improved design flexibility. Each routerless NoC size has an "optimal" wiring overlap that balances hardware overhead (to support additional paths/loops for data transfer) and performance [17]. As a result, the "optimal" loop configuration for a given set of constraints may vary substantially. A static heuristic-based design process such as REC [163] cannot take advantage of additional resources and, more importantly, cannot even generate designs with extremely strict constraints. The other prior approach, based on a genetic search, could explore these design constraints, but does not directly exploit past experiences and instead relies upon random combinations of past designs. DRL successfully optimizes around these constraints, yet adheres to a well-defined exploration framework revolving around Monte Carlo tree searches and strict action/reward guidelines. As a result, DRL reliably generates 4×4 NoCs in seconds and 10×10 NoCs in minutes.

The overall take-away is that DRL can provide a more holistic design strategy than heuristics due to improved flexibility in state/action/reward representation, yet remain efficient due to intelligent exploration and optimization. These combined capabilities provide a platform upon which future work can explore a variety of related architectural design challenges. As mentioned in the paper [17], the framework could be applied to more complex 3D NoC designs with relatively simple modifications. Additional generalizations could be made to explore underutilized wiring resources in silicon interposers, irregular or non-intuitive interconnect structures for chiplet networks, and high-performance NoCs for accelerators.

4.3   UNSUPERVISED LEARNING IN MEMORY SYSTEMS

Chapter 3 and the prior case studies showed that both supervised learning and RL can be applied to a wide variety of tasks and components, all while advancing the state of the art. Unsupervised learning, in comparison, may initially seem somewhat more limited since it does not accommodate human-defined goals in the way that training labels or a reward function allow. Some applications, however, benefit from this independence, resulting in highly efficient and practical solutions. In this case study, we highlight one such application based on work by Wang and Ipek [83]. Using k-majority clustering, a variant of k-means clustering (described in Chapter 2), they enable a low-overhead encoding scheme for data transmission that far surpasses both traditional coding schemes and recent heuristic-based methods.

4.3.1   BACKGROUND

Modern computer architecture designs necessitate significant data movement between memory and processing elements (e.g., CPUs, GPUs, domain-specific accelerators, etc.). This paradigm, coupled with high energy costs for data transmission,[1] has resulted in systems dominated by data transmission energy. In fact, recent work has shown that data movement energy in common tasks comprises approximately 50–80% of total system energy for consumer devices [169]. These high energy costs have motivated diverse design strategies for interconnects between memory,

---

[1]In a 10nm process, transmitting 256 bits of data across chip requires roughly $23\times$ the amount of energy as a 64-bit fused-multiply add. If this data is instead accessed from external DRAM, the data transmission energy cost is further increased by roughly two orders of magnitude [168].

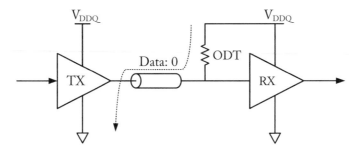

Figure 4.10: Termination with pseudo open drain (POD).

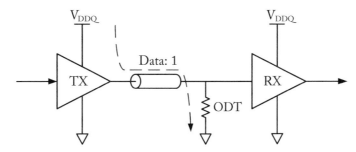

Figure 4.11: Termination with low voltage swing termination logic (LVSTL).

caches, and processing elements. In the following, we review important concepts and a few design strategies to better understand the benefits of data encoding.

Interconnect implementation, particularly the termination strategy, plays a fundamental role in determining how energy is consumed during data transmission. In *terminated interconnects*, a resistor is used to match the impedance of the transmission line. This resistor can be placed external to the chip or on-die, in which case it is referred to as an *on-die termination* (ODT). Many DRAM interfaces, including DDR3, DDR4, GDDR4, GDDR5, and LPDDR4, all support ODT in some manner, albeit with varying implementation strategies [83]. Specifically, DDR4 and GDDR5 adopt pseudo-open-drain (POD) signaling, illustrated in Figure 4.10. When the transmitted data is zero, there exists a static path (as shown by the dotted line) for current flow from V_{DDQ}, through the ODT, to ground. As a result, energy will be dissipated. In contrast, when the transmitted data is one, the RX input will be pulled high by the ODT, after which minimal energy will be dissipated. Opposite functionality is achieved through the use of low voltage swing terminated logic (LVSTL), depicted in Figure 4.11. This termination approach, adopted by LPDDR4, consumes energy when the transmitted data is one (shown by the dashed line). Conversely, when the transmitted data is zero (or encoded to be zero), the input to RX will be pulled to ground by the ODT, after which energy dissipation will be minimal.

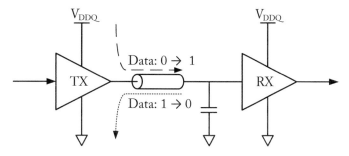

Figure 4.12: Unterminated transmission line with capacitor.

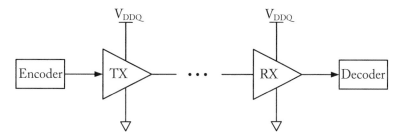

Figure 4.13: Generic transmit/receive setup with encoder/decoder.

Unterminated interconnects are another prevalent design strategy that exists in LPDDR3 and on-chip interconnects, such as those for caches [83]. This design, illustrated in Figure 4.12, forgoes an ODT and instead transmits data via charge stored on a capacitor. As shown by the dashed line, this capacitor must be charged when a transition from zero to one occurs and, as shown by the dotted line, must be discharged when a transition from one to zero occurs. Finally, if the data remains constant, then there will be minimal steady state energy dissipation.

4.3.2   MOTIVATION FOR MACHINE LEARNING

All of the aforementioned interconnect design strategies can be exploited to minimize energy consumption when particular data values are sent. This is achieved, as depicted in Figure 4.13, by encoding data prior to transmission and then decoding data following transmission. As such, a variety of coding methods have been proposed, some based on static guarantees and others based on dynamic behavior. As discussed below, these works do not fully exploit data similarity (both short-term and long-term) or introduce substantial overhead, both of which hinder optimal energy efficiency.

A simple approach used in DDR4 chips is data bus inversion (DBI) coding. Assuming POD termination, DBI coding can minimize energy consumption by guaranteeing that the number of transmitted 1s (i.e., the *Hamming weight*) is greater than $\frac{n}{2}$, where n is the number of bits. Prior to transmission, the number of 1s in the data is checked. If this number is less than $\frac{n}{2}$ then the data is inverted (i.e., 0s are replaced with 1s and vice versa). With this scheme, only a single additional bit is required to indicate to the decoder whether the original data was inverted. DBI coding can also be extended to unterminated interconnects by checking against the previously sent value and inverting the data if more than $n/2$ bits differ. As a whole, this scheme offers low overhead, but limited energy saving potential. When the data has close to an even number of ones and zero, there is practically no benefit since the inverted data would, again, have close to an even number of ones and zeros. More recent work (CAFO [170]) extends bus inversion to two dimensions and thereby achieves further improvements, particularly for memory endurance and reliability, but retains the same inherent limitations as DBI/BI.

Many works have instead explored heuristic-based methods that target similarities in transmitted data, potentially achieving greater energy reductions, albeit without strict guarantees. One straightforward method involves XORing the data to be transmitted with the previously transmitted data. Frequent value (FV) encoding instead [171] maintains a table of common data patterns that are, at runtime, compared with data to be transmitted. Upon a match, this data is replaced by a one-hot representation of the corresponding table identifier. If there is no match, the original data is instead sent. Combinations of XOR- and FV-based methods have also been explored [172].

Both XOR- and FV-based schemes, as well as extensions, rely upon similarities in *actual* data that has been or will be transmitted. In a situation where two dissimilar chunks of data are accessed in sequence, little benefit would be possible. The same is also true with DBI/BI and CAFO coding. An ideal approach would identify and exploit similarities, without these limitations, to maximize energy savings in diverse scenarios. This can be achieved through the use of machine-learning-based clustering.

4.3.3   IMPLEMENTATION

We now focus on the energy efficient coding scheme proposed by Wang and Ipek [83]. Their work is based on k-majority clustering, a variant of k-means clustering, and supported by an efficient low-level architectural implementation. We begin with an overview of k-majority clustering and then discuss architectural considerations.

K-majority Clustering

Recall from Chapter 2 that, in k-means clustering, each cluster centroid is placed to minimize the Euclidean (i.e., straight line) distance from that centroid to all data points belonging to that cluster. K-majority clustering is essentially the bitwise analog: data points are assigned to the cluster with minimal Hamming distance (i.e., the number of differing bits) and, subsequently, each centroid bit is set to one if the majority of data points in that cluster have a one for that specific bit. An example with three data points is depicted in Figure 4.14. Two out of three data

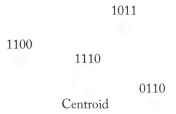

Figure 4.14: K-majority clustering example.

points have a 1 in the leftmost bit, thus the centroid also has a 1 for that bit. All other bits are calculated via the same procedure.

The algorithm for both k-means clustering and k-majority clustering follows an iterative procedure in which data points are relocated to the nearest cluster and then centroid placement is updated to better represent nearby points. This procedure is, by necessity, performed offline with all data points included. Data encoding, however, necessitates online updates to adapt to changes in data behavior, especially in a multi-workload environment. This change is possible using a saturating counter for each bit in every centroid. These centroids, although no longer precise, gain the ability to drift over time with changes in data. Most importantly, this approach does not require previous values to be stored, thus updates remain relatively simple. This modified algorithm (see Algorithm 4.5) updates the saturating counter for each centroid bit only when that centroid is the closest to the new data point. The majority vote for a specific centroid bit is then determined by the most significant bit (MSB) of its respective saturating counter.

Hardware Implementation and Design Trade-offs

Referring back to Figure 4.13, hardware implementation involves two primary components: the encoder block and the decoder block. For illustrative purposes, we study an implementation with two clusters. Cluster identification therefore requires one extra wire beyond that required for data transmission. As such, we show one byte of data sent per transmission to remain consistent with the 8:1 data-to-cluster wiring ratio specified in the paper [83].

Implementation for the encoder block is depicted in Figure 4.15. As specified by Algorithm 4.5 lines 1–4, incoming data is XORed with all cluster centroids to generate candidate encodings or residuals. These residuals are compared by the minimum Hamming weight selector, which outputs the ID of the most efficient coding. This ID, along with the selected residual (via the mux), is then transmitted along the interconnect. At the same time, both the original data and the cluster ID that generated the residual are sent to the updater block. Observe that these cluster updates are not on the critical path and therefore do not contribute to encoding latency.

Implementation for the decoder block, shown in Figure 4.16, is similar to that of the encoder, albeit without the need to compare residuals. Upon receiving the residual, the cluster

---

Algorithm 4.5 Pseudocode for online K-majority clustering algorithm.

Input : Set of m cluster centroids $\mathcal{K} = \{k_1, \ldots, k_m\}$ (assumed to be initialized previously), each with n bits

Input : New data point $\mathcal{X}$ with n bits

Input : Set of m by n cluster centroid counters $C = \{c_{00}, \ldots, c_{0n}, \ldots, c_{m0}, \ldots, c_{mn}\}$ (one per bit in each centroid)

1: **for all** clusters in $\mathcal{K}$ **do**
2:    Calculate the Hamming distance between data point $\mathcal{X}$ and centroid k_m
3: **end for**
4: Record the cluster ID m with the smallest Hamming distance
5: **for** $i \leftarrow 0$ to $n - 1$ **do**
6:    **if** bit i in data point $\mathcal{X}$, (i.e., $\mathcal{X}_i$) is one **then**
7:       Increment centroid counter c_{mi}
8:    **else**
9:       Decrement centroid counter c_{mi}
10:   **end if**
11:   Set bit i of cluster centroid k_m to the MSB of centroid counter c_{mi}
12: **end for**

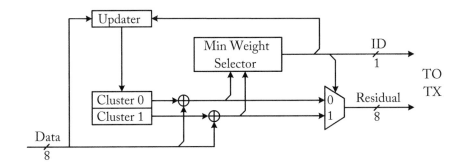

Figure 4.15: Encoder implementation for k-majority clustering (adapted from [83]).

ID is used to directly select the appropriate cluster center, after which the residual is XORed with the same cluster as at the encoder. The result is the original data which, as in the encoder block, is passed to the updater block along with the cluster ID.

This overall procedure requires that the clusters at both the encoder and decoder are always updated together in order to ensure correct transmission. Consequently, updates are only

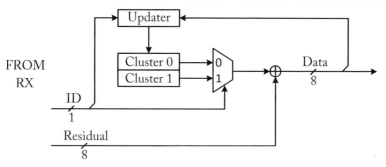

Figure 4.16: Decoder implementation for k-majority clustering (adapted from [83]).

performed after residuals have been calculated at the encoder and after the original data has been recovered at the decoder. Updates are then performed using hardware shown in Figure 4.17. As noted previously, each individual cluster is updated only when the original data is encoded/decoded using that cluster. Within the selected cluster updater, the original data is broken into bits (line 5 in Algorithm 4.5). Next, the saturating counter for each centroid bit is updated based on whether that bit in the original data was one or zero (line 6–10 in Algorithm 4.5). Finally, each centroid bit is updated by taking the MSB from each saturating counter (line 11 in Algorithm 4.5).

This hardware implementation, particularly the saturating counters, introduces several trade-offs between energy savings from efficient data encoding and overhead, in terms of both power and area.

- First is the number of centers. More centers increases the likelihood that a cluster with low Hamming weight, relative to the data to be transmitted, can be used for encoding; in fact, on average, the 32 most frequent data values represents roughly 32% of total transmitted data, and up to 68% in some applications [171]. Assuming data is transmitted in one-byte blocks, using two clusters fixed at all 1s and all 0s would be similar to DBI coding. Of course, since the clusters are not fixed, even two clusters could improve energy efficiency. More centers also, however, increases the number of bits required to encode the cluster ID and the number of saturating counters. Consequently, there will be a point of diminishing returns, after which overhead will grow faster than energy savings (discussed further in Section 4.3.4).

- Another significant trade-off involves the number of bits per saturating counter. As mentioned earlier, these saturating counters allow clusters to drift and adapt to changes in transmitted data. Increasing the counter size could help to stabilize centroid location and filter out noisy behavior since more updates would be required before the MSB is flipped and the centroid is updated. Too large a counter, however, would extend the

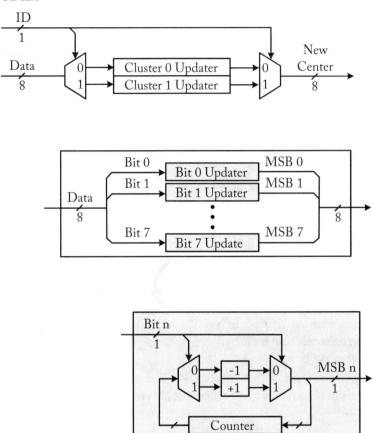

Figure 4.17: Updater implementation for k-majority clustering (adapted from [83]).

time for clusters to adjust to more significant changes in transmitted data. Of course, overhead would also increase proportionally with counter size.

• Another option to reduce overhead is by reducing center update frequency. Specifically, centers could be updated every X transmissions, both at the encoder and decoder. Similar to increased saturating counter size, a lower update frequency would introduce delay when adjusting cluster centers to new data behavior. Wang and Ipek [83] find that an interval of 64 transmissions achieves within 5% optimal energy saving due to encoding, while saving dynamic energy that would be required to increment/decrement counters and update centers.

4.3.4   RESULTS

Evaluations show that the cluster-based encoding scheme proposed by Wang and Ipek [83] reduces total system energy across a variety of applications and system configurations. These energy savings result from substantial reductions in the amount of transmitted 0s, both in DRAM and the LLC. Specifically, mobile systems observe a 40% reduction in transmitted 0s for both DRAM and LLC while server systems observe a 39% and 41% reduction in the DRAM and LLC, respectively, when compared with DBI coding. These improvements in coding efficiency are, in turn, made possible by clusters that better match the original data. As mentioned previously, machine-learning-based clustering does not require cluster centroids to be an *actual* data value, so they can instead be placed more centrally and reduce average Hamming distance compared to schemes that store previously transmitted data. Furthermore, this cluster-based scheme allows the choice between random cluster initialization (for the numbers above) or cluster placement based on offline workload analysis, which is shown to enable a further 9% reduction in transitions. This idea, while compatible with some prior work (e.g., frequent value encoding [171]), can be fully exploited with cluster-based encoding since dynamic updates can be made to account for slight variations in runtime behavior.

Overhead for hardware to support the proposed coding scheme is roughly comparable to prior work. Latency added by the encoding and decoding process results in 0.5% performance degradation compared to a system with DBI coding. Power and area overhead for the proposed encoder/decoder/updater is higher than prior work, but overall system energy is still lower due to more efficient data encodings.

Encoding efficiency in multi-application environments is another advantage for the proposed cluster-based work. Specifically, for a mobile system simultaneously running two SPEC2006 benchmarks, DRAM (LPDDR3) system energy is reduced by 13% (down from 15% in a single-workload environment) under the proposed encoding scheme, while the next best scheme provides only 4% reduction (down from 7% in a single-workload environment), both compared with the DBI baseline. The next best scheme, based on XORing with recent values, may have difficulty exploiting temporal locality when multiple data streams are interleaved. The proposed scheme, on the other hand, could allow individual clusters to be re-purposed by individual workloads based on data similarities, potentially reducing conflicts.

SUMMARY

The three case studies examined in this chapter serve as exemplars for machine learning application to architecture. The first case study on supervised learning branch predictors followed a series of developments leading to industry adoption. In particular, we observed how task-specific optimizations can enable perceptrons to generate predictions at the cycle level while integrating multiple source of information (i.e., perspectives), thereby achieving state-of-the-art accuracy. The second case study on design space exploration demonstrated how DRL can be applied to search immense spaces surpassing 10^{100}. Key optimizations based on multi-task learning for the ANN and MCTS-augmented search provided an efficient foundation for the framework. Further, the highly generalizable nature of the state, action, and reward function naturally lends itself to diverse follow-up works. Finally, the third case study on unsupervised learning in memory systems highlighted how machine learning algorithms can be adapted for hardware implementation. We also considered several important trade-offs relating to model accuracy and overhead; reducing the number of clusters (more generally, the model complexity), the number of bits for clusters (model weights), and the update frequency are all possible schemes that can be considered in future works.

In the next chapter, we move away from specific applications and instead analyze the range of techniques that have been applied in existing works.

CHAPTER 5

Analysis of Current Practice

The broad range of machine learning applications in architecture has produced an equally diverse array of strategies to select learning models, optimize implementation strategies, and adapt to task-specific needs. In this chapter, we revisit works introduced in Chapter 3 but, this time, we analyze varying techniques employed in these works. These comparisons identify potentially useful design practices and strategies for future work.

Work is divided into two categories that represent a natural division in design constraints and operating timescales, and therefore correspond to differing design practices. The first category, online machine learning applications, encompasses works that directly apply machine learning techniques at runtime, even if training is performed offline. Design complexity in this work is inherently limited by strict constraints on power, area, and real-time processing capability. The second category, offline machine learning applications, instead applies machine learning to guide architectural implementation, involving tasks such as design and simulation. Consequently, models for offline machine learning applications can exploit higher complexity and higher overhead options at the cost of training/prediction time.

5.1   ONLINE MACHINE LEARNING APPLICATIONS

5.1.1   MODEL AND FEATURE SELECTION

Model selection in online applications, although less flexible than in offline applications, still involves several choices that can affect the task setup and the resulting performance. Individual supervised learning predictions can be viewed as mutually independent, simply mapping distinct inputs to distinct outputs, while reinforcement learning involves a more continuous trajectory of actions that may build on each other, hence its common use in design space exploration. Note that applications for these two approaches are not necessarily disjoint, particularly for control tasks. In these tasks, a supervised learning approach may seek to predict the performance for a set of possible configurations and then select the configuration with the highest predicted performance. In contrast, a reinforcement learning approach could directly select configurations as actions. We can observe a direct comparison between these two approaches in work by Fettes et al. [97] on NoC DVFS. They found that both supervised and reinforcement learning models can perform well, but the RL model is more universally applicable since the energy/throughput trade-off can be tailored to application needs and does not require threshold tuning. Control at a system-level may also favor reinforcement learning in some tasks due to the built-in discount factor for rewards, which emphasizes the long-term significance of actions. For example, dynamic cache allocation between several workloads can be negatively impacted if cache ways are frequently swapped between the two workloads. In particular, recent work using deep Q-learning observed a tendency to select consistent configurations across many time-steps [134]. This certainly does not mean that RL is a one-model-fits-all solution. Supervised learning models find strong application in function approximation [136–138, 140] and branch prediction tasks [87, 88], which are far less suitable (if not impossible) to approach using RL since these tasks cannot be represented well as a sequence of actions.

Online machine learning applications tend to use either decision trees or ANNs, in the case of supervised learning models, and either Q-learning or deep Q-learning, in the case of RL models. As mentioned in Section 2.1.1, ANNs may perform better than decision trees for more complex tasks with high-dimension and continuous features. Similarly, deep Q-learning tends to be better suited for continuous-value actions. In both cases, these benefits have to be weighed against overhead (discussed in the next section). Unsupervised learning applications, such as that in the third case study, are still relatively uncommon and have focused on k-means clustering and variants. Consequently, the main challenge is devising a task setup that is amenable to unsupervised learning.

Finally, feature selection in online (and offline) applications is almost always performed offline to limit overhead. The overall goal is then to find a minimal feature set using techniques described in Section 2.1.3 that still meets model accuracy requirements.

5.1.2   IMPLEMENTATION AND OVERHEAD

Implementation for online machine learning applications faces limitations in data availability, storage space for models, etc., indicating the need for models that are both efficient and low complexity. These limitations will likely become more important to consider as more research moves toward real-world implementation.

Global data collection has proven to be one challenging aspect. In NoCs, for example, applications such as per-node injection throttling and hotspot prevention rely upon some degree of coordination between all the routers. As such, predictions or control actions typically require data from all nodes, regardless of whether or not there is a machine learning model implemented at every node or just at a centralized node. In some situations, this extra data to support machine learning models may negatively impact overall NoC latency or even lead to new hotspots, thus outweighing the benefits of any control actions. A number of approaches can be employed to resolve this problem, each offering a trade-off between the collection interval and overhead. First, we could use a separate, low-overhead network dedicated to data for the machine learning model(s). For instance, Daya et al. [102] used a separate network that propagates starvation data for each node across the network, allowing individual node-based Q-learning agents to determine appropriate injection throttling. Similarly, a dedicated network can be used to collect buffer utilization and virtual channel occupancy statistics [103]. A second approach still uses the original network, but limits the amount of traffic caused by machine learning data. Won et al. [119] encodes data into unused bits in standard packet headers and then opportunistically collects this data as packets pass through the router connected to a central control unit. This method introduces potential concerns about data staleness but, using sufficiently large control windows, performance can be nearly identical to omniscient data collection [173]. Smaller time windows can also be accommodated by sending dedicated packets [92]. Similar trade-offs when implementing either dedicated networks or more opportunistic schemes can be applied to more broad cases as well.

Implementation can also consider the use of either dedicated hardware for acceleration or software implementation using machine learning libraries. This dedicated hardware usually results in lower execution time, especially since fewer instructions may be required [136]. However, software-based implementation for simpler models such as classification trees can still achieve comparable performance to more complex models with dedicated hardware [140]. Consequently, specific tasks may exhibit practical trade-offs between area/power and accuracy/performance. An intermediate approach can implement a mix of dedicated hardware and software to provide even greater flexibility. For example, Won et al. [119] implemented an ANN in software using an on-die microcontroller. This implementation requires several orders of magnitude more cycles (15 K cycles for inference) than a fully hardware implementation. Further, this solution incurs marginal losses in energy savings due to longer configuration intervals, yet avoids extra hardware design, making it a practical choice in some applications.

Approaches for hardware implementation may also vary based on the task. A "standard" ANN implementation involves a finite state machine for control, an array of multiply-accumulate (MAC) units for calculation, a register array to load and store results, and a lookup-table-based activation function [92]. In this case, both the MAC array width and calculation precision can be adjusted to balance power/area and accuracy/speed. Regardless, even with a highly-parallelized approach, it may not be practical to use a standard digital-circuit-based implementation for applications that require predictions/actions on the order of 1–10 cycles. Those cases have necessitated specialized mixed signal designs, such as in the branch predictor design by St. Amant et al. [87]. They realized dot products in analog circuitry, leveraging transistor sizing and current summing to achieve a feasible overhead. Variance also exists in hardware for RL models. The "standard" Q-learning implementation requires a lookup table to store state-action values. Several works [15, 81] have instead used *CMAC* [174], replacing a potentially extensive Q-learning table with multiple coarse-grained overlapping tables. This approach also included hashing, using hashed state attributes to index the CMAC tables. Taken together, these two techniques balance generalization and overhead, although may introduce collisions/interference depending on the task. Further pipelining the hashing, CMAC table lookup, and calculation allows more possible action-values to be evaluated per cycle, resulting in a low-delay Q-learning implementation that is worth considering in other problems.

5.1.3   OPTIMIZATION

Online machine learning applications with online training benefit from adaptivity to runtime workload characteristics. Despite these benefits, low model accuracy can negatively impact system performance, most notably at the start of execution or during periods of high variance in workload characteristics. Adaptations to control and learning can be considered to avoid these detrimental impacts. For example, we can mitigate the impact of poor actions during exploration by introducing "shadow" operations [71]. These operations are low confidence actions suggested by the model that are still used in model updates but not executed by the system. Consequently, the model gains feedback on the goodness of the action without negatively impacting the system. Another common option leverages an additional (typically heuristic-based) model that performs initial control actions while a machine learning model is trained. This approach exploits the lower start-up delay for most heuristic models, thus preventing poor decisions early on during application execution. Once the machine learning model is trained, control actions can also be made using a hybrid combination based on error and consistency, allowing complementary control [119]. In the simplest case, checking the performance of a default configuration, as in [85], provides a guarantee that the machine learning model will not perform worse than the default, but can perform better.

In most works, machine learning models replace existing approaches (i.e., traditional models). Nevertheless, several recent works [75, 91] have demonstrated significant advantages by combining both traditional (non-machine-learning) and machine learning approaches. These improvements are derived from the orthogonal prediction/decision-making capabilities of the two approaches, thus enabling synergistic performance improvements. This method can also enable lower-cost machine learning applications by focusing on particular shortcomings in traditional approaches. Both recent works [75, 91] consider just branch prediction, thus significant opportunities exist to explore this potential co-design paradigm.

5.2   OFFLINE MACHINE LEARNING APPLICATIONS

5.2.1   MODEL AND FEATURE SELECTION

Offline machine learning applications generally exhibit substantial model/feature diversity since the model itself is not tied to a particular architecture. Design space exploration, in particular, can be approached using either iterative search methods that construct a design step-by-step or more direct approaches that predict performance for a full configuration. With iterative methods, machine learning approaches are usually advantageous as they can exploit learned knowledge about the design space. At one end of the spectrum, we can use machine learning as a starting point to augment traditional search algorithms [175]. At the other extreme, DRL serves as the primary search tool, while organizing learned information in a Monte Carlo search tree, resulting in a highly efficient search process even in an immense design space. System-level design space exploration has favored more standard supervised learning approaches [65, 111,

114]. Specific model choices vary, with linear [65, 111] and nonlinear [114] regression models, as well as random forests and neural networks [111] finding implementation. As in online machine learning applications, some tasks are naturally limited to supervised learning methods. Cross-architecture prediction is an exemplar since the goal involves mapping (i.e., approximating a function) from CPU-based features to GPU performance or power [29, 39, 62, 64, 67].

All these offline applications generally allow a wider range of features from more sources. Whereas online applications typically rely upon just hardware performance counters, offline applications have made use of application characteristics (e.g., instruction mix) or other features that are difficult to obtain in a real system (e.g., global age for packets in a NoC). It also becomes more feasible to process features, most commonly normalizing features to aid model performance. Techniques for feature selection still tend to follow those described in Section 2.1.3.

5.2.2   OPTIMIZATION

Fundamentally, the usefulness of machine learning models in offline applications is dependent on their performance and overhead relative to traditional design approaches. Put another way, a highly-accurate model is not very useful if it cannot be run on available computing resources. For that reason, a primary challenge in these offline applications is finding a balance between model accuracy, data efficiency, and computation overhead.

Ensemble methods are a particularly powerful approach that combine the predictions made by multiple models, thus offering greater predictive capabilities than any individual model. These methods have been applied in online machine learning applications [85], but primarily find practical use in offline machine learning applications as ensembles can be made arbitrarily large (relative to available computation resources). Regardless, simply scaling the ensemble size and using all predictions is not necessarily the best approach. For example, we could train many machine learning models and then generate an ensemble using a subset of those models that generalize best [65]. Furthermore, outlier detection is possible by filtering predictions whose performance and/or power predictions differ greatly from the closest configuration in training data. Alternatively, we could keep all the trained models since some may be very strong predictors in one application but weak predictors in another application [67]. In this case, prediction accuracy can be improved by selecting a subset of predictions closest to the median. This general notion of outlier filtering has broader implications as well; using an ensemble, we can estimate overall predictor confidence with calculable values such as the variance for all the predictions [60], thus providing some indication whether training data was sufficiently representative.

Sampling method optimization, while not unique to architecture tasks, is nevertheless important to consider in improving model accuracy. In some cases, it may be necessary to consider potential systematic biases in prediction models. Specifically, uniform random sampling may not adequately capture performance relationships in a non-uniform configuration space (as in cache configurations using powers of two for sizing) [114]. One solution involves a low-discrepancy sampling technique, *SOBOL* [176], to remove this systematic bias and prevent performance over-prediction for low-end configurations.

5.3   INTEGRATING DOMAIN KNOWLEDGE

The powerful relationship learning capabilities offered by machine learning algorithms enable black-box implementation in many tasks (i.e., without consideration for task-specific characteristics), but may fail to capitalize on additional domain knowledge that could improve interpretability or overall model performance. Additionally, in some applications, domain knowledge can help identify aberrant behaviors and improve overall model usefulness. These themes are highlighted in several specific works, but can be generally applicable for machine learning applied to architecture.

One approach uses mechanistic-empirical models, synthesizing a domain knowledge based mechanistic framework with empirical machine-learning-based learning for specific parameters. These models simplify implementation compared to purely mechanistic models [23], can avoid incorrect assumptions made in purely mechanistic models [113], and can offer higher accuracy than purely empirical models by avoiding overfitting [23]. Eyerman et al. [23] also demonstrated how these models can be used to construct CPI stacks, allowing meaningful alternative design comparisons.

Several works have also highlighted reasons to scrutinize model behavior, particularly when performance is lower than expected. Deng et al. [85], in their work predicting optimal NVM write strategies, presented a case for tuning machine learning models based on task specific domain knowledge. Following initial analysis, they discovered how a single configuration parameter (wear quota) can result in higher complexity and sub-optimal prediction accuracy for IPC and system energy, even with quadratic regression and gradient boosting models. Excluding wear quota from the configuration space, then later applying it to the predicted optimal configuration, provided a 2–6% improvement in prediction accuracy. Ardalani et al. [67] similarly examined inherent imperfections in their learning model for cross-platform performance prediction. Some predictions can be easy for learning models and hard for humans, representing an ideal scenario for machine learning applications; the converse can also be true. In both cases, machine learning applications are strengthened by considering task characteristics.

CHAPTER 6

Future Directions of AI for Architecture

This chapter synthesizes observations from Chapter 3 and analysis from Chapter 5 to identify opportunities and detail the need for future work. These opportunities may come at the model level, exploiting improved implementation strategies and learning capabilities, or at the application level, addressing the need for generalized tools or exploring altogether new areas.

6.1 INVESTIGATING MODELS AND ALGORITHMS

Existing works generally apply machine learning at a single time-scale or level of abstraction. These limitations motivate investigation into models and algorithms that capture the hierarchical nature of architecture, both in terms of system design and execution characteristics.

6.1.1 PERFORM PHASE-LEVEL PREDICTION

Application analysis using basic blocks [177] has long been a useful method for simulation, made possible by identifying unique and representative phases in program execution. Phase-level prediction offers analogous benefits in machine learning applications as it prevents any features or other characteristics from being lost when averaging over the entire application execution [33]. Perhaps more importantly, phase-level analysis can provide more data than application-level analysis, thus easing machine learning data requirements. A few recent works, in particular, have demonstrated promising results, with high accuracy for both performance prediction [62] as well as energy and reliability (lifetime) [85]. In general, most work [28, 65, 114] has not yet adopted phase-level prediction techniques (or does not explicitly mention their methodology). Specifically, future work can explore predictions for control and system reconfiguration based on these phase-level behaviors, rather than either static windows or application-level behaviors.

6.1.2 EXPLOIT NANOSECOND SCALE

Coarse-grained machine learning, used in many DVFS control schemes, provides significant benefits over standard control-theory based schemes, yet fine-grained control can provide even greater efficiency. Specifically, analysis by Bai et al. [120] indicated very rapid changes in energy consumption, on the order of 1000 instructions for some applications. Exploiting these brief intervals requires careful consideration for both the model and the algorithm. Future work may optimize existing algorithms such as experience sharing [178] and hybrid/tandem control [119], or consider approaches more suited for novel models (e.g., hierarchical models). These approaches could also enable additional nanosecond-scale co-optimization opportunities, such as dynamic LLC partitioning, to extract further efficiency gains.

6.1.3   EXPLORE HIERARCHICAL AND MULTI-AGENT MODELS

Application execution in computer systems naturally follows a hierarchical structure in which, at the top level, tasks are allocated to cores, then cores are assigned dynamic power and resource budgets (e.g., LLC space), and finally, at the bottom level, data/control packets are sent between cores and memory. Consequently, a single machine learning model may struggle to learn appropriate design/control strategies. Furthermore, in the case of reinforcement learning models, it can be exceedingly difficult to accurately assign rewards to specific low-level actions based on their impact on overall execution time, energy efficiency, etc. One promising approach in recent work is hierarchical models [179]. Hierarchical reinforcement learning models enable goal-directed learning that is particularly beneficial in environments with sparse feedback (e.g., task allocation). Applying hierarchical learning to architecture could therefore enable more effective multi-level design and control. Multi-agent models are another promising area in machine learning research. These models tend to focus on problems in which RL agents have only partial observability of their environment. Although partial-observability may not be a primary concern in individual computer systems, recent work [180] has applied this concept to internet packet routing and demonstrated convergence benefits via improved cooperation between individual agents.

6.2   ENHANCING IMPLEMENTATION STRATEGIES

Increasingly complex models require effective strategies and techniques to reduce overhead and enable practical implementation. Model pruning and weight quantization, as discussed below, are two particularly effective techniques with proven benefits in accelerators, while many other promising approaches are also being explored [181]. Our intention here is not to exhaustively present this line of research; instead, we highlight potential benefits from applying a few techniques, specifically in architectural applications.

6.2.1   APPLY MODEL PRUNING

Model complexity can be a limiting factor in online machine learning applications. In particular, a typical Q-learning approach requires a potentially extensive table to store action-values while neural-network-based approaches for both RL (in deep Q-networks) and supervised learning require low-latency storage for weights and may introduce substantial computational requirements. Consequently, neural network implementation in existing works is generally limited to a few layers, with many using just one hidden layer [92, 119, 129, 139] and some using one or two hidden layers [136, 137].

Recent research on neural networks has demonstrated promising methods to reduce model complexity through pruning [182, 183]. The general intuition is that many connections are unnecessary and can therefore be removed. Iteratively pruning a high-complexity network, then retraining from scratch on the sparse architecture achieves good results, with some work demonstrating very high sparsity (>90%) and little accuracy penalty [183].

Pruning applied to neural networks, either in deep Q-learning or supervised learning regression/classification, offers a method to train complex models for high accuracy, then prune for feasible implementation. Online machine learning applications using deep Q-learning have, thus far, been limited to a few works [97, 105, 134], one of which is currently impractical to implement [105]. Future work may instead consider pruned deep Q-networks as a useful alternative to standard Q-learning approaches. Pruning also provides a substantial opportunity for future work on performance prediction (as in DVFS control) and function approximation (as in machine learning enabled approximate computing). System-level approximation (discussed in Section 6.4) may particularly benefit from pruning high complexity models.

6.2.2   EXPLORE QUANTIZATION

Existing work primarily applies quantization to state values in Q-learning to enable practical Q-table implementation. Similarly, neural networks benefit from potential reduction in execution time, power, and area by reducing multiply-accumulator precision. Recent works, however, suggest a new spectrum of opportunities for alternative hardware implementations based on reduced precision models.

Binary neural networks, for example, quantize weights to be either +1 or −1, enabling computation based on binary bit-wise operations rather than decimal arithmetic operations [184]. An additional approach considered quantizing neural network weights into finite (but non-binary) subsets in order to replace multiply operations with lookup-table accesses [185], allowing higher precision and lower execution time, albeit with potentially higher area cost. Future work on machine learning applications can exploit similar hardware implementations while exploring optimal quantization levels for various tasks and control schemes.

6.3   DEVELOPING GENERALIZED TOOLS

Existing machine learning tools (e.g., scikit-learn [186]) have proven useful for relatively simple machine learning applications. Nevertheless, complex design and simulation tasks require more sophisticated tools to enable rapid task-specific optimizations using general-purpose frameworks.

6.3.1   ENABLE BROAD APPLICATION AND OPTIMIZATION

Purpose-built architectural tools, similar to heuristic design strategies, can be useful in enabling design, exploration, and simulation that satisfies a common use case. These approaches may still be limited in their application to a specific problem, optimization criteria, system configuration, etc. Given the fast-paced nature of architectural research (and machine learning research), there is a need to develop more generalized tools and easily modifiable frameworks to address broader applications and optimization options.

Machine-learning-based design tools are especially promising, with recent works demonstrating successful application to immense design spaces. Opportunities for new design tools are not, however, limited to specific architectural components. Chip layout is a notable example in which even simple clustering algorithms can substantially outperform existing heuristic approaches [48]. Future work can also continue to develop more broadly applicable tools for performance and power prediction. In particular, recent work on cross-platform performance prediction [68] suggests the possibility for high prediction accuracy with purely static features, thus representing another potential area for additional research.

6.3.2   ENCOURAGE WIDESPREAD USAGE

Future widespread applications of machine learning in architecture will likely be dependent upon the accessibility of design tools, especially for individuals without *a priori* machine learning experience. Depending upon the task, the current process of machine learning application generally requires testing a number of models, data processing and representation methods, search algorithms, etc. More generalized frameworks could potentially help to simplify this process by pre-configuring some aspects (e.g., hyperparameters) that may benefit from expert knowledge. With this foundation, task-specific application may involve simply gathering training data (in a supervised learning setting) or defining action/reward representation (in an RL setting). For example, recent work [17] envisioned reuse of a DRL framework for diverse NoC-related design tasks involving interposers, chiplets, and accelerators. Although the framework might not be fully compatible with all work, especially in novel areas, other aspects such as state representation may remain compatible, thus providing a stronger starting point for further applications, especially for individuals with limited machine learning background.

6.4   EMBRACING NOVEL APPLICATIONS

Opportunities abound for future work to apply machine learning to both existing and emerging architectures, replace traditional approaches to enable long-term scaling, and advance capabilities for automated design.

6.4.1   EXPEDITE EMERGING TECHNOLOGIES

Several proposals [78, 84–86] establish how machine learning can be used to optimize both standard (energy, performance) and non-standard (lifetime, tail-latency) criteria. These non-standard criteria are shown to be particularly problematic in emerging technologies as these technologies cannot easily find widespread implementation without some guarantees, particularly for total lifetime. Therefore, applying machine learning to optimize both standard and non-standard criteria can help future work to intelligently recognize signs of reliability issues and proactively adjust control strategies to maintain critical guarantees. Future work can continue to explore both performance and reliability concerns in both non-volatile memories and other technologies (e.g., spin-transfer-torque RAM).

Machine learning applications are also motivated by the lack of best-practice knowledge when implementing these emerging technologies in real-world designs. In general, work in long-standing design areas, such as task allocation and branch prediction, may easily incorporate best-practice domain knowledge to guide approaches, whether applying machine learning or some other traditional methods. In contrast, best practices for emerging technologies may not be immediately obvious. Consequently, traditional methods may perform poorly. Machine learning models, on the other hand, can learn directly from data, thereby obviating the need for human expertise in some applications. For example, machine learning application to 2D photonic NoCs [94], 2.5D processing-in-memory designs [70], and 3D NoCs [107–109] have all shown strong performance over existing approaches. Future work can explore machine learning application to novel concerns such as connectivity and reconfigurability in interposers and domain-specific accelerators. Applications in novel, cross-disciplinary areas, such as those involving brain-machine implants (e.g., Bhattacharjee [187]), can similarly benefit from machine-learning-based prediction and control while integrating best practice knowledge from existing domains.

6.4.2   EXPAND SYSTEM-LEVEL APPROXIMATE COMPUTING

As discussed in Section 3.6, machine learning applications for approximate computing have been mostly limited to function approximation. There are, however, many other facets of approximate computing that have already been implemented in non-machine-learning works and can reap additional benefits by utilizing machine learning. For example, APPROX-NoC [188] reduces network traffic using approximated data, while encoding/decoding is performed using frequent pattern matching. Encoding/decoding could instead be combined with a scheme similar to that in the third case study (Section 4.3), perhaps focusing approximation on bits that would incur a transition. Another work explored a multi-faceted approximation scheme for a smart camera system [189] using approximate DRAM (lower refresh rate), approximate algorithms (loop skipping) and approximate data (lower sensor resolution). Approximations in the existing work are based on hill-climbing, thus representing an ideal scenario for efficient machine-learning-based search. Compiler-based work [190] for system-wide approximation has improved prior capabilities to determine approximable code, but relies upon heuristic searches with representative inputs. Consequently, this method does not provide strong statistical guarantees, such as those in MITHRA [139]. Future work may explore searches based on DRL (or perhaps hierarchical reinforcement learning) to incorporate existing approximation techniques into a scalable framework for high-dimensional approximation and co-optimization.

6.4.3   EXPLORE VALUE PREDICTION

Value prediction was originally proposed in the 1990s as a way to bypass data dependencies by speculatively executing instructions using predicted values. At the time, IPC gains were generally considered to be outweighed by additional processor complexity [191]. Regardless, a number of works [192–195] based on the recently proposed EOLE [191] microarchitecture have generated renewed interest in practical value prediction.

Approaches for value prediction, similarly to branch prediction, fall within two categories [196]: computational predictors, which perform an operation on previous values to produce new predictions, and context-based predictors, which correlate recent value history and previously recorded value history. Most recent works, including the current state-of-the-art [193], consider a combination of these two approaches, thus allowing individual predictors to focus on particular patterns. Given the success of machine learning in branch prediction, we view machine-learning-based value prediction as another potentially promising area that, to our knowledge, has not been considered by existing works. In particular, future work could consider applying machine learning models to improve accuracy on patterns that are difficult to predict using more traditional contextual-based methods. Additional practical benefits could be achieved by optimizing predictions for particular workloads (as was suggested for processor cluster-gating in [125]).

6.4.4   IMPLEMENT SYSTEM-WIDE, COMPONENT-LEVEL OPTIMIZATION

Recent work has begun to explore broader machine-learning-based design and optimization strategies. MLNoC [111] explores a wide SoC feature space for NoC design optimization. Core and uncore DVFS are combined in machine learned machines [130], along with LLC dynamic cache partitioning to explore co-optimization potential at runtime. Related DNN accelerator research [197] proposed co-optimization of hardware-based (e.g., bitwidth) and neural network parameters (e.g., L2 regularization). These works motivate consideration for system-wide, component-level machine learning applications.

Existing system-level optimization schemes (e.g., [124, 131, 198]) consider configuration opportunities at just a single and very high level of abstraction (e.g., task allocation or big.LITTLE core configurations). Specifically, we observe a disconnect between overall system optimization goals and lower-level considerations in sub-systems, such as NoC packet routing or cache prefetching, which could negatively affect performance or reliability. Although these works may still include some related features such as NoC utilization or DRAM bandwidth in their machine learning models, we see further opportunities with system-wide, component-level frameworks for runtime optimization. In such a framework, control decisions would involve a larger hierarchy of both component-level (or lower) features and control options as well as higher-level decisions, allowing a more comprehensive and precise perspective for runtime optimization.

6.4.5 ADVANCE AUTOMATED DESIGN

While fully automated design might be the ultimate objective, increasingly automated design is nevertheless an important milestone for work in the near future. Specifically, as more tasks are automated, there is greater potential to enable a positive-feedback loop between machine learning and architecture, providing immense practical benefits for both fields. There are, of course, a number of intervening challenges that must be solved, each of which represents a substantial area for future work.

One challenge involves modeling the hierarchical structure of architectural components. This model would likely benefit from integrating pertinent characteristics across the system stack, from process technology to full-system behavior, thus generating a highly accurate representation for real-world systems. Another research direction could explore methods for machine learning models to help automatically identify potential design aspects for improvement. Ideally, this model could explore not just reconfiguration of pre-existing options (as in [199]), but also generate novel configuration options. Integrating these and potentially other capabilities may provide a framework to significantly advance automated design.

CHAPTER 7

Conclusions

Starting from a few pioneering works, AI-based architectural design have expanded and now stand at the forefront of a potentially new paradigm.

AI-based approaches are a natural fit for some of the most challenging problems in architecture design. Problems involving performance/power prediction, or practically any other criteria, can leverage vast amounts of data and identify underlying relationships that may be unintuitive (or even impossible) for a human to identify. Similarly, exploration in vast design spaces is made possible via models that automatically learn the solution-space structure and prioritize promising regions. Dynamic control can also be enhanced by models that proactively predict, rather than react, to changes in demands while considering the long-term benefits of various actions. These capabilities are supported by a diverse range of training methods and learning models, thereby enabling task-specific optimizations in numerous scenarios.

This broad applicability, coupled with fundamental limitations in traditional design methods, has inspired significant work exploring AI-based architecture design. In fact, machine learning has already been successfully applied to practically all major components, including the core, cache, NoC, and memory, with performance often surpassing prior state-of-the-art analytical, heuristic, and human-expert strategies. The effectiveness of these strategies is highlighted by our case studies. In the first case study, we saw how a simple perceptron model can achieve state-of-the-art accuracy in branch prediction. The second case study on DRL demonstrates unparalleled search capabilities even under strict design constraints. Finally, the third case study indicates how unsupervised learning can provide practical solutions, even without human guidance. These applications are likely just the beginning of a revolutionary shift in architecture.

Optimization opportunities at the model level involving pruning and quantization offer broad benefits by enabling more practical implementations. Similarly, opportunities abound to extend existing work using ever-more-powerful machine learning models, enabling finer granularity, system-wide implementation. Finally, machine learning may be applied to entirely new aspects of architecture, learning hierarchical, or abstract representations to characterize full system behaviors based on both high- and low-level details. These extensive opportunities, along with yet to be envisioned possibilities, may close the loop on highly (or even fully) automated architectural design.

Bibliography

[1] F. Rosenblatt, The perceptron: A probabilistic model for information storage and organization in the brain, in *Psychological Review*, 65(6), 1958. DOI: 10.1037/h0042519. 1

[2] M. Minsky and S. A. Papert, *Perceptrons: An Introduction to Computational Geometry*, MIT Press, 1969. DOI: 10.7551/mitpress/11301.001.0001. 1

[3] D. Silver, J. Schrittwieser, K. Simonyan, I. Antonoglou, A. Huang, A. Guez, T. Hubert, L. Baker, M. Lai, A. Bolton, Y. Chen, T. Lillicrap, F. Hui, L. Sifre, G. van den Driessche, T. Graepel, and D. Hassabis, Mastering the game of Go without human knowledge, in *Nature*, 550(7676):354–359, 2017. DOI: 10.1038/nature24270. 1, 7, 72, 74

[4] O. Russakovsky, J. Deng, H. Su, J. Krause, S. Satheesh, S. Ma, Z. Huang, A. Karpathy, A. Khosla, M. Bernstein, A. C. Berg, and L. Fei-Fei, ImageNet large scale visual recognition challenge, in *International Journal of Computer Vision (IJCV)*, 115(3):211–252, 2015. DOI: 10.1007/s11263-015-0816-y. 2

[5] Y. LeCun, L. Bottou, Y. Bengio, and P. Haffner, Gradient-based learning applied to document recognition, in *Proc. of the IEEE Machine Learning Research*, 86(11), 1998. DOI: 10.1109/5.726791. 2

[6] A. Krizhevsky, I. Sutskever, and G. Hinton, ImageNet classification with deep convolutional neural networks, in *Advances in Neural Information Processing Systems 25*, pages 1097–1105, Curran Associates, Inc., 2012. DOI: 10.1145/3065386. 2

[7] K. Simonyan and A. Zisserman, Very deep convolutional networks for large-scale image recognition, *ArXiv:1409.1556*, 2014. 2

[8] C. Szegedy, W. Liu, Y. Jia, P. Sermanet, S. Reed, D. Anguelov, D. Erhan, V. Vanhoucke, and A. Rabinovich, Going deeper with convolutions, *ArXiv:1409.4842*, 2014. DOI: 10.1109/cvpr.2015.7298594. 2

[9] K. He, X. Zhang, S. Ren, and J. Sun, Deep residual learning for image recognition, *ArXiv:1512.03385*, 2015. DOI: 10.1109/cvpr.2016.90. 2

[10] Q. Xie, M.-T. Luong, E. Hovy, and Q. V. Le, Self-training with noisy student improves ImageNet classification, *ArXiv:1911.04252*, 2019. DOI: 10.1109/cvpr42600.2020.01070. 2

[11] Computer History Museum, Timeline of computer history, 2020. https://www.computerhistory.org/timeline/ 1

[12] G. E. Moore, Cramming more components onto integrated circuits, in *Electronics*, 38(8), 1965. DOI: 10.1109/n-ssc.2006.4785860. 1

[13] R. H. Dennard, F. H. Gaensslen, H. Yu, V. L. Rideout, E. Bassous, and A. R. LeBlanc, Design of ion-implanted MOSFETs with very small physical dimensions, in *IEEE Journal of Solid-State Circuits*, 9(5), 1974. DOI: 10.1109/jssc.1974.1050511. 2

[14] D. A. Jiménez and C. Lin, Dynamic branch prediction with perceptrons, in *International Symposium on High-Performance Computer Architecture (HPCA)*, January 2001. DOI: 10.1109/hpca.2001.903263. 2, 46, 48, 64, 67

[15] E. Ipek, O. Mutlu, J. F. Martinez, and R. Caruana, Self-optimizing memory controllers: A reinforcement learning approach, in *International Symposium on High-Performance Computer Architecture (HPCA)*, June 2008. DOI: 10.1109/isca.2008.21. 2, 45, 48, 89

[16] M. Ozsoy, K. N. Khasawneh, C. Donovick, I. Gorelik, N. Abu-Ghazaleh, and D. Ponomarev, Hardware-based malware detection using low-level architectural features, in *IEEE Transactions on Computers*, vol. 65, March 2016. DOI: 10.1109/tc.2016.2540634. 2, 58, 59

[17] T.-R. Lin, D. Penney, M. Pedram, and L. Chen, A deep reinforcement learning framework for architectural exploration: A routerless NoC case study, in *International Symposium on High-Performance Computer Architecture (HPCA)*, February 2020. DOI: 10.1109/hpca47549.2020.00018. 3, 50, 53, 69, 70, 71, 72, 74, 76, 77, 96

[18] D. DiTomaso, T. Boraten, A. Kodi, and A. Louri, Dynamic error mitigation in NoCs using intelligent prediction techniques, in *International Symposium on Microarchitecture (MICRO)*, October 2016. DOI: 10.1109/micro.2016.7783734. 3, 51, 54

[19] A. Seznec, TAGE-SC-L branch predictors again, in *5th JILP Workshop on Computer Architecture Competitions (JWAC-5): Championship Branch Prediction (CBP-5)*, June 2016. 6, 46, 68

[20] D. A. Jiménez, Multiperspective perceptron predictor, in *5th JILP Workshop on Computer Architecture Competitions (JWAC-5): Championship Branch Prediction (CBP-5)*, June 2016. 6, 68

[21] A. Seznec, A 256 Kbits L-TAGE branch predictor, in *Journal of Instruction-Level Parallelism (JILP) Special Issue: The Second Championship Branch Prediction Competition (CBP-2)*, vol. 9, May 2007. 6, 46, 68

[22] N. Mishra, J. D. Lafferty, and H. Hoffman, CALOREE: Learning control for predictable latency and low energy, in *International Conference on Architectural Support for Programming Languages and Operating Systems (ASPLOS)*, March 2018. DOI: 10.1145/3173162.3173184. 6, 55, 57

[23] S. Eyerman, K. Hoste, and L. Eeckhout, Mechanistic-empirical processor performance modeling for constructing CPI stacks on real hardware, in *International Symposium on Performance Analysis of Systems and Software (ISPASS)*, April 2011. DOI: 10.1109/ispass.2011.5762738. 6, 40, 41, 92

[24] J. Tromp and G. Farnebäck, Combinatorics of Go, in *Computers and Games*, pages 84–99, Springer Berlin Heidelberg, 2007. DOI: 10.1007/978-3-540-75538-8_8. 7

[25] C. Young, Using TPUs to design TPUs, in *International Workshop on AI-Assisted Design for Architecture (AIDArc)*, held in conjunction with ISCA, 2018. 7

[26] S. Kotsiantis, Supervised machine learning: A review of classification techniques, in *Proc. of the Conference on Emerging Artificial Intelligence Applications in Computer Engineering: Real World AI Systems with Applications in eHealth, HCI, Information Retrieval and Pervasive Technologies*, pages 3–24, 2007. 9, 19

[27] T. A. Stephenson, An introduction to Bayesian network theory and usage, *Tech. Rep., IDIAP*, 2000. 11

[28] N. Mishra, H. Zhang, J. D. Lafferty, and H. Hoffman, A probabilistic graphical model-based approach for minimizing energy under performance constraints, in *International Conference on Architectural Support for Programming Languages and Operating Systems (ASPLOS)*, March 2015. DOI: 10.1145/2694344.2694373. 12, 55, 57, 93

[29] I. Baldini, S. J. Fink, and E. Altman, Predicting GPU performance from CPU runs using machine learning, in *International Symposium on Computer Architecture and High Performance Computing (SBAC-PAD)*, October 2014. DOI: 10.1109/sbac-pad.2014.30. 15, 42, 44, 91

[30] A. Smola and B. Schölkopf, A tutorial on support vector regression, in *Statistics and Computing*, 14:199–222, 2004. DOI: 10.1023/b:stco.0000035301.49549.88. 16

[31] V. N. Vapnik, An overview of statistical learning theory, in *IEEE Transactions on Neural Networks*, vol. 10, September 1999. DOI: 10.1109/72.788640. 16

[32] B. C. Csáji, Approximation with artificial neural networks, Master's thesis, Eötvös Loránd University, 2001. 17

[33] Y. Li, D. Penney, A. Ramamurthy, and L. Chen, Characterizing on-chip traffic patterns in general-purpose GPUs: A deep learning approach, in *International Conference on Computer Design (ICCD)*, November 2019. DOI: 10.1109/iccd46524.2019.00016. 18, 28, 43, 44, 93

[34] J. Chung, C. Gulcehre, K. Cho, and Y. Bengio, Empirical evaluation of gated recurrent neural networks on sequence modeling, *ArXiv:1412.3555*, December 2014. 18

[35] A. Bordes, S. Ertekin, J. Weston, and L. Bottou, Fast kernel classifiers with online and active learning, in *Machine Learning Research*, 6:1579–1619, 2005. 19

[36] I. Guyon and A. Elisseeff, An introduction to variable and feature selection, in *The Journal of Machine Learning Research*, 3:1157–1182, March 2003. 23, 24

[37] J. Li, K. Chen, S. Wang, F. Morstatter, R. P. Trevino, J. Tang, and H. Liu, Feature selection: A data perspective, in *ACM Computing Surveys*, vol. 50, January 2018. DOI: 10.1145/3136625. 24

[38] J. Shlens, A tutorial on principal component analysis, 2014. *ArXiv:1404.1100*, 2014. 24

[39] X. Zheng, P. Ravikumar, L. K. John, and A. Gerstlauer, Learning-based analytical cross-platform performance prediction, in *International Conference on Embedded Computer Systems: Architectures, Modeling, and Simulation (SAMOS)*, July 2015. DOI: 10.1109/samos.2015.7363659. 24, 40, 41, 91

[40] J. Clark and F. Provost, Unsupervised dimensionality reduction vs. supervised regularization for classification from sparse data, in *Data Mining and Knowledge Discovery*, vol. 33, February 2019. DOI: 10.1007/s10618-019-00616-4. 25

[41] C. R. Shalizi, *Advanced Data Anlysis from an Elementary Point of View*, Cambridge University Press, 2017. 25, 26

[42] I. T. Jolliffe, *Principal Components Analysis*, New York, Springer-Verlag, 2002. DOI: 10.1007/b98835. 25, 26

[43] P. Virtanen, R. Gommers, T. E. Oliphant, M. Haberland, T. Reddy, D. Cournapeau, E. Burovski, P. Peterson, W. Weckesser, J. Bright, S. J. van der Walt, M. Brett, J. Wilson, K. Jarrod Millman, N. Mayorov, A. R. J. Nelson, E. Jones, R. Kern, E. Larson, C. Carey, İ. Polat, Y. Feng, E. W. Moore, J. Vand erPlas, D. Laxalde, J. Perktold, R. Cimrman, I. Henriksen, E. A. Quintero, C. R. Harris, A. M. Archibald, A. H. Ribeiro, F. Pedregosa, P. van Mulbregt, and SciPy 1.0 Contributors, SciPy 1.0: Fundamental algorithms for scientific computing in Python, in *Nature Methods*, 17:261–272, 2020. DOI: 10.1038/s41592-019-0686-2. 26

[44] Y. Lu, I. Cohen, X. S. Zhou, and Q. Tian, Feature selection using principal feature analysis, in *ACM International Conference on Multimedia*, September 2007. DOI: 10.1145/1291233.1291297. 27

[45] C. Ding and X. He, K-means clustering via principal component analysis, in *International Conference on Machine Learning (ICML)*, July 2004. DOI: 10.1145/1015330.1015408. 27

[46] G. Wu, J. L. Greathouse, A. Lyashevsky, N. Jayasena, and D. Chiou, GPGPU performance and power estimation using machine learning, in *International Symposium on High-Performance Computer Architecture (HPCA)*, February 2015. DOI: 10.1109/hpca.2015.7056063. 27, 41, 42, 44, 54

[47] L. van der Maaten and G. Hinton, Visualizing data using t-SNE, in *Machine Learning Research*, 9:2579–2605, 2008. 27, 28

[48] G. Wu, Y. Xu, D. Wu, M. Ragupathy, Y. Yen Mo, and C. Chu, Flip-flop clustering by weighted k-means algorithm, in *Design Automation Conference (DAC)*, June 2016. DOI: 10.1145/2897937.2898025. 28, 96

[49] M. Alawieh, F. Wang, and X. Li, Efficient hierarchical performance modeling for integrated circuits via Bayesian co-learning, in *Design Automation Conference (DAC)*, June 2017. DOI: 10.1145/3061639.3062235. 29

[50] S. Yu, B. Krishnapuram, R. Rosales, and R. B. Rao, Bayesian co-training, in *Machine Learning Research*, 12:2649–2680, 2011. 29

[51] H. Lakkaraju, E. Kamar, R. Caruana, and E. Horvitz, Identifying unknown unknowns in the open world: Representations and policies for guided exploration, in *Conference on Artificial Intelligence*, February 2017. 30

[52] L. P. Kaelbling, M. L. Littman, and A. W. Moore, Reinforcement learning: A survey, May 1996. *ArXiv:cs/9605103*. DOI: 10.1613/jair.301. 32

[53] C. J. C. H. Watkins and P. Dayan, Q-learning, in *Machine Learning*, vol. 8, May 1989. DOI: 10.4018/978-1-59140-555-9.ch114. 32, 33

[54] G. Rummery and M. Niranjan, On-line Q-learning using connectionist systems, in *Technical Report CUED/F-INFENG/TR 166*, November 1994. 33

[55] R. S. Sutton and A. G. Barto, *Reinforcement Learning: An Introduction*, 2nd ed., Cambridge, MIT Press, 1998. DOI: 10.1109/tnn.1998.712192. 34

[56] R. J. Williams, Simple statistical gradient-following algorithms for connectionist reinforcement learning, in *Machine Learning*, vol. 8, May 1992. DOI: 10.1007/bf00992696. 34

[57] G. Hamerly, E. Perelman, J. Lau, and B. Calder, SimPoint 3.0: Faster and more flexible program phase analysis, in *Journal of Instruction Level Parallelism*, vol. 7, 2005. 39, 41

[58] M. Badr and N. Enright Jerger, SynFull: Synthetic traffic models capturing cache coherent behaviour, in *International Symposium on Computer Architecture (ISCA)*, June 2014. DOI: 10.1109/isca.2014.6853236. 39, 41

[59] J. Yin, O. Kayiran, M. Poremba, N. Enright Jerger, and G. H. Loh, Efficient synthetic traffic models for large, complex SoCs, in *International Symposium on High-Performance Computer Architecture (HPCA)*, March 2016. DOI: 10.1109/hpca.2016.7446073. 40, 41

[60] E. Ipek, S. A. McKee, B. R. de Supinski, M. Schulz, and R. Caruana, Efficiently exploring architectural design spaces via predictive modeling, in *International Conference on Architectural Support for Programming Languages and Operating Systems (ASPLOS)*, October 2006. DOI: 10.1145/1168919.1168882. 40, 41, 91

[61] B. Ozisikyilmaz, G. Memik, and A. Choudhary, Machine learning models to predict performance of computer system design alternatives, in *International Conference on Parallel Processing (ICPP)*, September 2008. DOI: 10.1109/icpp.2008.36. 40, 41

[62] X. Zheng, L. K. John, and A. Gerstlauer, Accurate phase-level cross-platform power and performance estimation, in *Design Automation Conference (DAC)*, June 2016. DOI: 10.1145/2897937.2897977. 40, 41, 91, 93

[63] N. Agarwal, T. Jain, and M. Zahran, Performance prediction for multi-threaded applications, in *International Workshop on AI-assisted Design for Architecture (AIDArc)*, held in conjunction with ISCA, June 2019. 40, 41

[64] W. Jia, K. A. Shaw, and M. Martonosi, Stargazer: Automated regression-based GPU design space exploration, in *International Symposium on Performance Analysis of Systems and Software (ISPASS)*, April 2012. DOI: 10.1109/ispass.2012.6189201. 41, 42, 44, 91

[65] A. Jooya, N. Dimopoulos, and A. Baniasadi, Multiobjective GPU design space exploration optimization, in *International Conference on High Performance Computing and Simulation (HPCS)*, July 2016. DOI: 10.1109/hpcsim.2016.7568398. 41, 42, 44, 90, 91, 93

[66] T.-R. Lin, Y. Li, M. Pedram, and L. Chen, Design space exploration of memory controller placement in throughput processors with deep learning, in *IEEE Computer Architecture Letters*, vol. 18, March 2019. DOI: 10.1109/lca.2019.2905587. 42, 44

[67] N. Ardalani, C. Lestourgeon, K. Sankaralingam, and X. Zhu, Cross-architecture performance prediction (XAPP) using CPU code to predict GPU performance, in *International Symposium on Microarchitecture (MICRO)*, June 2015. DOI: 10.1145/2830772.2830780. 42, 44, 91, 92

[68] N. Ardalani, U. Thakker, A. Albarghouthi, and K. Sankaralingam, A static analysis-based cross-architecture performance prediction using machine learning, in *International Workshop on AI-assisted Design for Architecture (AIDArc)*, held in conjunction with ISCA, June 2019. 42, 44, 96

[69] K. O'Neal, P. Brisk, E. Shriver, and M. Kishinevsky, HALWPE: Hardware-assisted light weight performance estimation for GPUs, in *Design Automation Conference (DAC)*, June 2017. DOI: 10.1145/3061639.3062257. 42, 44

[70] A. Pattnaik, X. Tang, A. Jog, O. Kayıran, A. K. Mishra, M. T. Kandemir, O. Mutlu, and C. R. Das, Scheduling techniques for GPU architectures with processing-in-memory capabilities, in *International Conference on Parallel Architectures and Compilation Techniques (PACT)*, September 2016. DOI: 10.1145/2967938.2967940. 43, 44, 97

[71] L. Peled, S. Mannor, U. Weiser, and Y. Etsion, Semantic locality and context-based prefetching using reinforcement learning, in *International Symposium on High-Performance Computer Architecture (HPCA)*, June 2015. DOI: 10.1145/2749469.2749473. 43, 47, 90

[72] Y. Zeng and X. Guo, Long short term memory based hardware prefetcher, in *International Symposium on Memory Systems (MemSys)*, October 2017. DOI: 10.1145/3132402.3132405. 43, 47

[73] M. Hashemi, K. Swersky, J. A. Smith, G. Ayers, H. Litz, J. Chang, C. E. Kozyrakis, and P. Ranganathan, Learning memory access patterns, in *International Conference on Machine Learning (ICML)*, July 2018. 43, 47

[74] P. Braun and H. Litz, Understanding memory access patterns for prefetching, in *International Workshop on AI-assisted Design for Architecture (AIDArc)*, held in conjunction with ISCA, June 2019. 43, 47

[75] E. Bhatia, G. Chacon, S. Pugsley, E. Teran, P. V. Gratz, and D. A. Jiménez, Perceptron-based prefetch filtering, in *International Symposium on Computer Architecture (ISCA)*, June 2019. DOI: 10.1145/3307650.3322207. 43, 47, 90

[76] J. Hiebel, L. E. Brown, and Z. Wang, Machine learning for fine-grained hardware prefetcher control, in *International Conference on Parallel Processing (ICPP)*, August 2019. DOI: 10.1145/3337821.3337854. 45, 47

[77] E. Teran, Z. Wang, and D. A. Jiménez, Perceptron learning for reuse prediction, in *International Symposium on Microarchitecture (MICRO)*, October 2016. DOI: 10.1109/micro.2016.7783705. 45, 47

[78] H. Wang, X. Yi, P. Huang, B. Cheng, and K. Zhou, Efficient SSD caching by avoiding unnecessary writes using machine learning, in *International Conference on Parallel Processing (ICPP)*, August 2018. DOI: 10.1145/3225058.3225126. 45, 47, 97

[79] A. Margaritov, D. Ustiugov, E. Bugnion, and B. Grot, Virtual address translation via learned page tables indexes, in *Conference on Neural Information Processing Systems (NeurIPS)*, December 2018. 45, 47

[80] T. Kraska, A. Beutel, E. H. Chi, J. Dean, and N. Polyzotis, The case for learned index structures, in *International Conference on Management of Data (SIGMOD)*, June 2018. DOI: 10.1145/3183713.3196909. 45

[81] J. Mukundan and J. F. Martinez, MORSE: Multi-objective reconfigurable self-optimizing memory scheduler, in *International Symposium on High-Performance Computer Architecture (HPCA)*, February 2012. DOI: 10.1109/hpca.2012.6168945. 45, 48, 89

[82] S. Manoj, H. Yu, H. Huang, and D. Xu, A Q-learning based self-adaptive I/O communication for 2.5D integrated many-core microprocessor and memory, in *IEEE Transactions on Computers*, vol. 65, June 2015. DOI: 10.1109/tc.2015.2439255. 45, 48

[83] S. Wang and E. Ipek, Reducing data movement energy via online data clustering and encoding, in *International Symposium on Microarchitecture (MICRO)*, October 2016. DOI: 10.1109/micro.2016.7783735. 45, 48, 77, 78, 79, 80, 81, 82, 83, 84, 85

[84] W. Kang and S. Yoo, Dynamic management of key states for reinforcement learning-assisted garbage collection to reduce long tail latency in SSD, in *Design Automation Conference (DAC)*, June 2018. DOI: 10.1109/dac.2018.8465934. 46, 48, 97

[85] Z. Deng, L. Zhang, N. Mishra, H. Hoffman, and F. T. Chong, Memory cocktail therapy: A general learning-based framework to optimize dynamic tradeoffs NVMs, in *International Symposium on Microarchitecture (MICRO)*, October 2017. DOI: 10.1145/3123939.3124548. 46, 48, 90, 91, 92, 93, 97

[86] J. Xiao, Z. Xiong, S. Wu, Y. Yi, H. Jin, and K. Hu, Disk failure prediction in data centers via online learning, in *International Conference on Parallel Processing (ICPP)*, June 2018. DOI: 10.1145/3225058.3225106. 46, 48, 97

[87] R. St. Amant, D. A. Jiménez, and D. Burger, Low-power, high-performance analog neural branch prediction, in *International Symposium on Microarchitecture (MICRO)*, November 2008. DOI: 10.1109/MICRO.2008.4771812. 46, 48, 88, 89

[88] D. A. Jiménez, An optimized scaled neural branch predictor, in *International Conference on Computer Design (ICCD)*, October 2011. DOI: 10.1109/iccd.2011.6081385. 46, 48, 88

[89] E. Garza, S. Mirbagher-Ajorpaz, T. A. Khan, and D. A. Jiménez, Bit-level perceptron prediction for indirect branches, in *International Symposium on Computer Architecture
(ISCA)*, June 2019. DOI: 10.1145/3307650.3322217. 46, 48

[90] A. Seznec, A 64-Kbytes ITTAGE indirect branch predictor, in *2nd JILP Workshop
on Computer Architecture Competitions (JWAC-2): Championship Branch Prediction*, June
2011. 46

[91] S. J. Tarsa, C.-K. Lin, G. Keskin, G. Chinya, and H. Wang, Improving branch prediction
by modeling global history with convolutional neural networks, in *International Workshop
on AI-assisted Design for Architecture (AIDArc)*, held in conjunction with ISCA, June 2019.
46, 48, 90

[92] A. G. Savva, T. Theocharides, and V. Soteriou, Intelligent on/off dynamic link management for on-chip networks, in *Journal of Electrical and Computer Engineering – Special issue
on Networks-on-Chip: Architectures, Design Methodologies, and Case Studies*, January 2012.
DOI: 10.1155/2012/107821. 49, 52, 53, 89, 95

[93] D. DiTomaso, A. Sikder, A. Kodi, and A. Louri, Machine learning enabled power-aware
network-on-chip design, in *Design, Automation and Test in Europe (DATE)*, March 2017.
DOI: 10.23919/date.2017.7927203. 49, 52, 53

[94] S. V. Winkle, A. Kodi, R. Bunescu, and A. Louri, Extending the power-efficiency
and performance of photonic interconnects for heterogeneous multicores with machine
learning, in *International Symposium on High-Performance Computer Architecture (HPCA)*,
February 2018. DOI: 10.1109/HPCA.2018.00048. 49, 52, 53, 97

[95] M. F. Reza, T. T. Le, B. De, M. Bayoumi, and D. Zhao, Neuro-NoC: Energy optimization in heterogeneous many-core NoC using neural networks in dark silicon era, in
International Symposium on Circuits and Systems (ISCAS), May 2018. DOI: 10.1109/is-
cas.2018.8351580. 49, 52, 53

[96] M. Clark, A. Kodi, R. Bunescu, and A. Louri, LEAD: Learning-enabled energy-aware
dynamic voltage/frequency scaling in NoCs, in *Design Automation Conference (DAC)*,
June 2018. DOI: 10.1109/dac.2018.8465925. 49, 52, 53

[97] Q. Fettes, M. Clark, R. Bunescu, A. Karanth, and A. Louri, Dynamic voltage and frequency scaling in NoCs with supervised and reinforcement learning techniques, in *IEEE
Transactions on Computers*, vol. 68, March 2019. DOI: 10.1109/tc.2018.2875476. 49, 52,
53, 87, 95

[98] J. A. Boyan and M. L. Littman, Packet routing in dynamically changing networks: A
reinforcement learning approach, in *Advances in Neural Information Processing Systems*,
6:671–678, 1994. 49, 53

[99] M. Majer, C. Bobda, A. Ahmadinia, and J. Teich, Packet routing in dynamically changing networks on chip, in *International Parallel and Distributed Processing Symposium (IPDPS)*, April 2005. DOI: 10.1109/ipdps.2005.323. 49, 53

[100] C. Feng, Z. Lu, A. Jantsch, J. Li, and M. zhang, A reconfigurable fault-tolerant deflection routing algorithm based on reinforcement learning for network-on-chip, in *International Workshop on Network on Chip Architectures (NoCArc)*, held in conjunction with MICRO, December 2010. DOI: 10.1145/1921249.1921254. 49, 53

[101] M. Ebrahimi, M. Daneshtalab, and F. Farahnakian, HARAQ: Congestion-aware learning model for highly adaptive routing algorithm in on-chip networks, in *International Symposium on Networks-on-Chip (NOCS)*, June 2012. DOI: 10.1109/nocs.2012.10. 49, 53

[102] B. K. Daya, L.-S. Peh, and A. P. Chandrakasan, Quest for high-performance bufferless NoCs with single-cycle express paths and self-learning throttling, in *Design Automation Conference (DAC)*, June 2016. DOI: 10.1145/2897937.2898075. 49, 53, 88

[103] V. Soteriou, T. Theocharides, and E. Kakoulli, A holistic approach towards intelligent hotspot prevention in network-on-chip-based multicores, in *IEEE Transactions on Computers*, vol. 65, May 2015. DOI: 10.1109/tc.2015.2435748. 50, 53, 88

[104] B. Wang, Z. Lu, and S. Chen, ANN based admission control for on-chip networks, in *Design Automation Conference (DAC)*, June 2019. DOI: 10.1145/3316781.3317772. 50, 53

[105] J. Yin, Y. Eckert, S. Che, M. Oskin, and G. H. Loh, Toward more efficient NoC arbitration: A deep reinforcement learning approach, in *International Workshop on AI-Assisted Design for Architecture (AIDArc)*, held in conjunction with ISCA, June 2018. 50, 53, 95

[106] J. Yin, S. Sethumurugan, Y. Eckert, C. Patel, A. Smith, E. Morton, M. Oskin, N. Enright Jerger, and G. H. Loh, Experiences with ML-driven design: A NoC case study, in *IEEE International Symposium on High Performance Computer Architecture (HPCA)*, February 2020. DOI: 10.1109/hpca47549.2020.00058. 50, 53

[107] S. Das, J. R. Doppa, D. H. Kim, P. P. Pande, and K. Chakrabarty, Optimizing 3D NoC design for energy efficiency: A machine learning approach, in *International Conference on Computer-Aided Design (ICCAD)*, November 2015. DOI: 10.1109/iccad.2015.7372639. 50, 53, 97

[108] S. Das, J. R. Doppa, P. P. Pande, and K. Chakrabarty, Energy-efficient and reliable 3D network-on-chip (NoC): Architectures and optimization algorithms, in *International Conference on Computer-Aided Design (ICCAD)*, November 2016. DOI: 10.1145/2966986.2980096. 50, 53, 97

[109] B. K. Joardar, R. G. Kim, J. R. Doppa, P. P. Pande, D. Marculescu, and R. Marculescu, Learning-based application-agnostic 3D NoC design for heterogeneous manycore systems, in *IEEE Transactions on Computers*, vol. 68, June 2019. DOI: 10.1109/tc.2018.2889053. 50, 53, 71, 97

[110] T.-R. Lin, D. Penney, M. Pedram, and L. Chen, Optimizing routerless network-on-chip designs: An innovative learning-based framework, *ArXiv:1905.04423*, May 2019. 50

[111] N. Rao, A. Ramachandran, and A. Shah, MLNoC: A machine learning based approach to NoC design, in *International Symposium on Computer Architecture and High Performance Computing (SBAC-PAD)*, September 2018. DOI: 10.1109/cahpc.2018.8645914. 50, 53, 90, 91, 98

[112] S. Bandyopadhyay, S. Saha, U. Maulik, and K. Deb, A simulated annealing-based multi-objective optimization algorithm: AMOSA, in *IEEE Transactions on Evolutionary Computation*, vol. 12, May 2008. DOI: 10.1109/tevc.2007.900837. 51, 71

[113] Z. Qian, D.-C. Juan, P. Bogdan, C.-Y. Tsui, D. Marculescu, and R. Marculescu, SVR-NoC: A performance analysis tool for network-on-chips using learning-based support vector regression model, in *Design, Automation and Test in Europe (DATE)*, March 2013. DOI: 10.7873/date.2013.083. 51, 53, 92

[114] K. Sangaiah, M. Hempstead, and B. Taskin, Uncore RPD: Rapid design space exploration of the uncore via regression modeling, in *International Conference on Computer-Aided Design (ICCAD)*, November 2015. DOI: 10.1109/iccad.2015.7372593. 51, 53, 91, 93

[115] K. Wang, A. Louri, A. Karanth, and R. Bunescu, High-performance, energy-efficient, fault-tolerant network-on-chip design using reinforcement learning, in *Design, Automation and Test in Europe (DATE)*, March 2019. DOI: 10.23919/date.2019.8714869. 51, 54

[116] K. Wang, A. Louri, A. Karanth, and R. Bunescu, IntelliNoC: A holistic design framework for energy-efficient and reliable on-chip communication for many-cores, in *International Symposium on Computer Architecture (ISCA)*, June 2019. DOI: 10.1145/3307650.3322274. 51, 54

[117] P. E. Bailey, D. K. Lowenthal, V. Ravi, B. Rountree, M. Schulz, and B. R. de Supinski, Adaptive configuration selection for power-constrained heterogeneous systems, in *International Conference on Parallel Processing (ICPP)*, September 2014. DOI: 10.1109/icpp.2014.46. 54, 57

[118] G.-Y. Pan, J.-Y. Jou, and B.-C. Lai, Scalable power management using multilevel reinforcement learning for multiprocessors, in *ACM Transactions on Design Automation of Electronic Systems*, August 2014. DOI: 10.1145/2629486. 54, 57

[119] J.-Y. Won, X. Chen, P. Gratz, J. Hu, and V. Soteriou, Up by their bootstraps: Online learning in artificial neural networks for CMP uncore power management, in *International Symposium on High-Performance Computer Architecture (HPCA)*, February 2014. DOI: 10.1109/hpca.2014.6835941. 54, 57, 89, 90, 94, 95

[120] Y. Bai, V. W. Lee, and E. Ipek, Voltage regulator efficiency aware power management, in *International Conference on Architectural Support for Programming Languages and Operating Systems (ASPLOS)*, April 2017. DOI: 10.1145/3037697.3037717. 54, 55, 57, 93

[121] M. Allen and P. Fritzsche, Reinforcement learning with adaptive Kanerva coding for Xpilot game AI, in *IEEE Congress of Evolutionary Computation*, June 2011. DOI: 10.1109/cec.2011.5949796. 54

[122] Z. Chen and D. Marculescu, Distributed reinforcement learning for power limited many-core system performance optimization, in *Design, Automation and Test in Europe (DATE)*, March 2015. DOI: 10.7873/date.2015.0992. 55, 57

[123] Z. Chen, D. Stamoulis, and D. Marculescu, Profit: Priority and power/performance optimization for many-core systems, in *IEEE Transactions on Computer-Aided Design of Integrated Circuits and Systems*, 37:2064–2075, October 2018. 55, 57

[124] C. Imes, S. Hofmeyr, and H. Hoffman, Energy-efficient application resource scheduling using machine learning classifiers, in *International Conference on Parallel Processing (ICPP)*, August 2018. DOI: 10.1145/3225058.3225088. 55, 57, 98

[125] S. J. Tarsa, R. B. R. Chowdhury, J. Sebot, G. Chinya, J. Gaur, K. Sankaranarayanan, C.-K. Lin, R. Chappell, R. Singhal, and H. Wang, Post-silicon CPU adaptation made practical using machine learning, in *International Symposium on Computer Architecture (ISCA)*, June 2019. DOI: 10.1145/3307650.3322267. 55, 57, 98

[126] D. Lo, T. Song, and G. E. Suh, Prediction-guided performance-energy trade-off for interactive applications, in *International Symposium on Microarchitecture (MICRO)*, December 2015. DOI: 10.1145/2830772.2830776. 55, 57

[127] S. J. Lu, R. Tessier, and W. Burleson, Reinforcement learning for thermal-aware many-core task allocation, in *Proc. of the 25th Ed. on Great Lakes Symposium on VLSI*, May 2015. DOI: 10.1145/2742060.2742078. 55, 58

[128] H. Zhang, B. Tang, X. Geng, and H. Ma, Learning driven parallelization for large-scale video workload in hybrid CPU-GPU cluster, in *International Conference on Parallel Processing (ICPP)*, August 2018. DOI: 10.1145/3225058.3225070. 55, 58

[129] R. Bitirgen, E. Ipek, and J. F. Martinez, Coordinated management of multiple interacting resources in chip multiprocessors: A machine learning approach, in *International Symposium on Microarchitecture (MICRO)*, November 2008. DOI: 10.1109/micro.2008.4771801. 56, 58, 95

[130] R. Jain, P. R. Panda, and S. Subramoney, Machine learned machines: Adaptive co-optimization of caches, cores, and on-chip network, in *Design, Automation and Test in Europe (DATE)*, March 2016. DOI: 10.3850/9783981537079_0083. 56, 58, 98

[131] D. Nemirovsky, T. Arkose, N. Markovic, M. Nemirovsky, O. Unsal, and A. Cristal, A machine learning approach for performance prediction and scheduling on heterogeneous CPUs, in *International Symposium on Computer Architecture and High Performance Computing (SBAC-PAD)*, October 2017. DOI: 10.1109/sbac-pad.2017.23. 56, 58, 98

[132] K. Ma, X. Li, S. R. Srinivasa, Y. Liu, J. Sampson, Y. Xie, and V. Narayanan, Spendthrift: Machine learning based resource and frequency scaling for ambient energy harvesting nonvolatile processors, in *Asia and South Pacific Design Automation Conference (ASP-DAC)*, January 2017. DOI: 10.1109/aspdac.2017.7858402. 56, 58

[133] R. Nishtala, P. Carpenter, V. Petrucci, and X. Martorell, Hipster: Hybrid task manager for latency-critical cloud workloads, in *International Symposium on High-Performance Computer Architecture (HPCA)*, February 2017. DOI: 10.1109/hpca.2017.13. 56, 58

[134] R. Nishtala, V. Petrucci, P. Carpenter, and M. Sjalander, Twig: Multi-agent task management for colocated latency-critical cloud services, in *International Symposium on High Performance Computer Architecture (HPCA)*, February 2020. DOI: 10.1109/hpca47549.2020.00023. 56, 58, 88, 95

[135] Y. Ding, N. Mishra, and H. Hoffmann, Generative and multi-phase learning for computer systems optimization, in *International Symposium on Computer Architecture (ISCA)*, June 2019. DOI: 10.1145/3307650.3326633. 56, 58

[136] H. Esmaeilzadeh, A. Sampson, L. Ceze, and D. Burger, Neural acceleration for general-purpose approximate programs, in *International Symposium on Microarchitecture (MICRO)*, December 2012. DOI: 10.1109/micro.2012.48. 59, 60, 88, 89, 95

[137] A. Yazdanbakhsh, J. Park, H. Sharma, P. Lotfi-Kamran, and H. Esmaeilzadeh, Neural acceleration for GPU throughput processors, in *International Symposium on Microarchitecture (MICRO)*, December 2015. DOI: 10.1145/2830772.2830810. 59, 60, 88, 95

[138] B. Grigorian, N. Farahpour, and G. Reinman, BRAINIAC: Bringing reliable accuracy into neurally-implemented approximate computing, in *International Symposium on High-Performance Computer Architecture (HPCA)*, February 2015. DOI: 10.1109/hpca.2015.7056067. 59, 60, 88

[139] D. Mahajan, A. Yazdanbaksh, J. Park, B. Thwaites, and H. Esmaeilzadeh, Towards statistical guarantees in controlling quality tradeoffs for approximate acceleration, in *International Symposium on Computer Architecture (ISCA)*, June 2016. DOI: 10.1109/isca.2016.16. 59, 60, 95, 98

[140] G. F. Oliveira, L. R. Goncalves, M. Brandalero, A. C. S. Beck, and L. Carro, Employing classification-based algorithms for general-purpose approximate computing, in *Design Automation Conference (DAC)*, June 2018. DOI: 10.1109/dac.2018.8465822. 59, 60, 88, 89

[141] F. N. Taher, J. Callenes-Sloan, and B. C. Schafer, A machine learning based hard fault recuperation model for approximate hardware accelerators, in *Design Automation Conference (DAC)*, June 2018. DOI: 10.1109/dac.2018.8465859. 59, 60

[142] T.-Y. Yeh and Y. N. Patt, Two-level adaptive training branch prediction, in *Proc. of the 24th ACM/IEEE International Symposium on Microarchitecture*, pages 51–61, November 1991. DOI: 10.1145/123465.123475. 63

[143] B. Burgess, Samsung's Exynos-M1 CPU, in *Hot Chips: A Symposium on High Performance Chips*, August 2016. 64

[144] J. Rupley, Samsung's Exynos-M3 CPU, in *Hot Chips: A Symposium on High Performance Chips*, August 2018. 64

[145] M. Shah, R. Golla, G. Grohoski, P. Jordan, J. Barreh, J. Brooks, M. Greenberg, G. Levinsky, M. Luttrell, C. Olson, Z. Samoail, M. Smittle, and T. Ziaja, Sparc T4: A dynamically threaded server-on-a-chip, in *IEEE Micro*, 32(2):8–19, 2012. DOI: 10.1109/mm.2012.1. 64

[146] C. Williams, "Neural network" spotted deep inside Samsung's Galaxy S7 silicon brain, in *The Register*, August 2016. 64

[147] F. Rosenblatt, *Principles of Neurodynamics: Perceptrons and the Theory of Brain Mechanisms*, Spartan, 1962. 64

[148] H. D. Block, The perceptron: A model for brain functioning, in *Reviews of Modern Physics*, 34:123–135, 1962. DOI: 10.1103/revmodphys.34.123. 64

[149] D. A. Jiménez and C. Lin, Neural methods for dynamic branch prediction, in *ACM Transactions on Computer Systems*, 20:369–397, November 2002. DOI: 10.1145/571637.571639. 64, 66

[150] T. H. Cormen, C. E. Leiserson, and R. L. Rivest, *Introduction to Algorithms*, McGraw Hill, 1990. DOI: 10.2307/2583667. 67

[151] D. A. Jiménez, S. W. Keckler, and C. Lin, The impact of delay on the design of branch predictors, in *Proc. of the 33rd Annual International Symposium on Microarchitecture (MICRO-33)*, pages 67–76, December 2000. DOI: 10.1109/micro.2000.898059. 67

[152] D. A. Jiménez, Reconsidering complex branch predictors, in *Proc. of the 9th International Symposium on High Performance Computer Architecture (HPCA-9)*, pages 43–52, February 2002. DOI: 10.1109/hpca.2003.1183523. 67

[153] D. A. Jiménez, Fast path-based neural branch prediction, in *Proc. of the 36th Annual IEEE/ACM International Symposium on Microarchitecture, MICRO 36*, page 243, IEEE Computer Society, Washington, DC, 2003. DOI: 10.1109/micro.2003.1253199. 67

[154] D. A. Jiménez, Piecewise linear branch prediction, in *Proc. of the 32nd Annual International Symposium on Computer Architecture (ISCA-32)*, June 2005. DOI: 10.1109/isca.2005.40. 67, 68

[155] D. Tarjan and K. Skadron, Merging path and gshare indexing in perceptron branch prediction, in *ACM Transactions on Archit. Code Optim.*, 2:280–300, September 2005. DOI: 10.1145/1089008.1089011. 67

[156] G. H. Loh and D. A. Jiménez, Reducing the power and complexity of path-based neural branch prediction, in *Proc. of the Workshop on Complexity-Effective Design (WCED'05)*, pages 28–35, June 2005. 67

[157] A. Seznec, Genesis of the O-GEHL branch predictor, in *Journal of Instruction-Level Parallelism (JILP)*, vol. 7, April 2005. 67

[158] The Journal of Instruction-Level Parallelism, *The Journal of Instruction-Level Parallelism 5th JILP Workshop on Computer Architecture Competitions (JWAC-5): Championship Branch Prediction*, http://www.jilp.org/cbp2016, June 2016. 68

[159] B. Grayson, J. Rupley, G. Z. Jr., E. Quinnell, D. A. Jiménez, T. Nakra, P. Kitchin, R. Hensley, E. Brekelbaum, V. Sinha, and A. Ghiya, Evolution of the Samsung Exynos CPU microarchitecture, in *47th ACM/IEEE Annual International Symposium on Computer Architecture, ISCA*, Valencia, Spain, May/June 2020. DOI: 10.1109/isca45697.2020.00015. 68

[160] P. Gratz, C. Kim, R. McDonald, S. W. Keckler, and D. Burger, Implementation and evaluation of on-chip network architecture, in *International Conference on Computer Design*, November 2007. DOI: 10.1109/iccd.2006.4380859. 70

[161] Y. Hoskote, S. Vangal, A. Singh, H. Borkar, and S. Borkar, A 5-GHz mesh interconnect for a teraflops processor, in *IEEE Micro*, November 2007. DOI: 10.1109/mm.2007.4378783. 70

[162] L. Chen and T. M. Pinkston, NoRD: Node-router decoupling for effective power-gating of on-chip routers, in *International Symposium on Microarchitecture*, December 2012. DOI: 10.1109/micro.2012.33. 70

[163] F. Alazemi, A. Azizimazreah, B. Bose, and L. Chen, Routerless networks-on-chip, in *IEEE International Symposium on High Performance Computer Architecture*, February 2018. DOI: 10.1109/hpca.2018.00049. 70, 71, 76, 77

[164] S. Liu, T. Chen, L. Li, X. Feng, Z. Xu, H. Chen, F. Chong, and Y. Chen, IMR: High-performance low-cost multi-ring NoCs, in *IEEE Transactions on Parallel and Distributed Systems*, vol. 27, June 2016. DOI: 10.1109/tpds.2015.2465905. 71

[165] C. Wu, C. Deng, L. Liu, J. Han, J. Chen, S. Yin, and S. Wei, A multi-objective model oriented mapping approach for NoC-based computing systems, in *IEEE Transactions on Parallel and Distributed Systems*, July 2016. DOI: 10.1109/tpds.2016.2589934. 71

[166] R. Caruana, Multitask learning, in *Machine Learning*, 28:41–75, 1997. DOI: 10.1023/A:1007379606734. 74

[167] C. Rosin, Multi-armed bandits with episode context, in *Annals of Mathematics and Artificial Intelligence*, 61:203–230, September 2010. DOI: 10.1007/s10472-011-9258-6. 75

[168] S. W. Keckler, W. J. Dally, B. Khailany, M. Garland, and D. Glasco, GPUs and the future of parallel computing, in *IEEE Micro*, 31(5):7–17, 2011. DOI: 10.1109/mm.2011.89. 77

[169] A. Boroumand, S. Ghose, Y. Kim, R. Ausavarungnirun, E. Shiu, R. Thakur, D. Kim, A. Kuusela, A. Knies, P. Ranganathan, and O. Mutlu, Google workloads for consumer devices: Mitigating data movement bottlenecks, in *International Conference on Architectural Support for Programming Languages and Operating Systems (ASPLOS)*, March 2018. DOI: 10.1145/3173162.3173177. 77

[170] R. Maddah, S. M. Seyedzadeh, and R. Melhem, CAFO: Cost aware flip optimization for asymmetric memories, in *International Symposium on High Performance Computer Architecture (HPCA)*, February 2015. DOI: 10.1109/hpca.2015.7056043. 80

[171] J. Yang, R. Gupta, and C. Zhang, Frequent value encoding for low power data buses, in *ACM Transactions on Design Automation of Electronic Systems*, 9:354–384, July 2004. DOI: 10.1145/1013948.1013953. 80, 83, 85

[172] S. Komatsu, M. Ikeda, and K. Asada, Low power chip interface based on bus data encoding with adaptive code-book method, in *Proc. 9th Great Lakes Symposium on VLSI*, March 1999. DOI: 10.1109/glsv.1999.757458. 80

[173] X. Chen, Z. Xu, H. Kim, P. Gratz, J. Hu, M. Kishinevsky, and U. Ogras, In-network monitoring and control policy for DVFS of CMP networks-on-chip and last level caches, in *International Symposium on Networks-on-Chip (NOCS)*, May 2012. DOI: 10.1109/nocs.2012.12. 89

[174] R. Sutton, Generalization in reinforcement learning: Successful examples using sparse coarse coding, in *Conference on Neural Information Processing Systems (NeurIPS)*, June 1996. 89

[175] J. A. Boyan and A. W. Moore, Learning evaluation functions to improve optimization by local search, in *The Journal of Machine Learning Research*, September 2001. 90

[176] P. Bratley and B. L. Fox, Algorithm 659:Implementing Sobol's quasirandom sequence generator, in *ACM Transactions on Mathematical Software*, vol. 14, March 1988. DOI: 10.1145/42288.214372. 92

[177] T. Sherwood, E. Perelman, and B. Calder, Basic block distribution analysis to find periodic behavior and simulation points in applications, in *International Conference on Parallel Architectures and Compilation Techniques (PACT)*, September 2001. DOI: 10.1109/pact.2001.953283. 93

[178] R. M. Kretchmar, Reinforcement learning algorithms for homogenous multi-agent systems, in *Workshop on Agent and Swarm Programming*, 2003. 94

[179] T. D. Kulkarni, K. R. Narasimhan, A. Saeedi, and J. B. Tenenbaum, Hierarchical deep reinforcement learning: Integrating temporal abstraction and intrinsic motivation, in *Conference on Neural Information Processing Systems (NeurIPS)*, December 2016. 94

[180] H. Mao, Z. Gong, Z. Zhang, Z. Xiao, and Y. Ni, Learning multi-agent communication under limited-bandwidth restriction for internet packet routing, *ArXiv:1903.05561*, February 2019. 94

[181] V. Sze, Y.-H. Chen, T.-J. Yang, and J. Emer, Efficient processing of deep neural networks: A tutorial and survey, *ArXiv:1703.09039*, August 2017. DOI: 10.1109/jproc.2017.2761740. 94

[182] S. Han, J. Pool, J. Tran, and W. J. Dally, Learning both weights and connections for efficient neural networks, *ArXiv:1506.02626*, October 2015. 95

[183] D. C. Mocanu, E. Mocanu, P. Stone, P. H. Nguyen, M. Gibescu, and A. Liotta, Scalable training of artificial neural networks with adaptive sparse connectivity inspired by network science, in *Nature Communications*, vol. 9, June 2018. DOI: 10.1038/s41467-018-04316-3. 95

[184] M. Courbariaux, I. Hubara, D. Soudry, R. El-Yaniv, and Y. Bengio, Binarized neural networks: Training deep neural networks with weights and activations constrained to +1 or -1, *ArXiv:1602.02830*, March 2016. 95

[185] M. S. Razlighi, M. Imani, F. Koushanfar, and T. Rosing, LookNN: Neural network with no multiplication, in *Design, Automation and Test in Europe (DATE)*, March 2017. DOI: 10.23919/date.2017.7927280. 95

[186] F. Pedregosa, G. Varoquaux, A. Gramfort, V. Michel, B. Thirion, O. Grisel, M. Blondel, P. Prettenhofer, R. Weiss, V. Dubourg, J. Vanderplas, A. Passos, D. Cournapeau, M. Brucher, M. Perrot, and E. Duchesnay, Scikit-learn: Machine learning in Python, in *Journal of Machine Learning Research*, 12:2825–2830, 2011. DOI: 10.1145/2786984.2786995. 95

[187] A. Bhattacharjee, Using branch predictors to predict brain activity in brain-machine implants, in *International Symposium on Microarchitecture (MICRO)*, October 2017. DOI: 10.1145/3123939.3123943. 97

[188] R. Boyapati, J. Huang, P. Majumder, K. H. Yum, and E. J. Kim, APPROX-NoC: A data approximation framework for network-on-chip architectures, in *International Symposium on Computer Architecture (ISCA)*, June 2017. DOI: 10.1145/3079856.3080241. 97

[189] A. Raha and V. Raghunathan, Towards full-system energy-accuracy tradeoffs: A case study of an approximate smart camera system, in *Design Automation Conference (DAC)*, June 2017. DOI: 10.1145/3061639.3062333. 97

[190] A. Sampson, A. Baixo, B. Ransford, T. Moreau, J. Yip, L. Ceze, and M. Oskin, AC-CEPT: A programmer-guided compiler framework for practical approximate computing, in *University of Washington Technical Report*, vol. 1, January 2015. 97

[191] A. Perais and A. Seznec, EOLE: Paving the way for an effective implementation of value prediction, in *International Symposium on Computer Architecture (ISCA)*, June 2014. DOI: 10.1109/isca.2014.6853205. 98

[192] R. Sheikh, H. W. Cain, and R. Damodaran, Load value prediction via path-based address prediction: Avoiding mispredictions due to conflicting stores, in *International Symposium on Microarchitecture (MICRO)*, October 2017. DOI: 10.1145/3123939.3123951. 98

[193] A. Seznec, Exploring value prediction with the eves predictor, in *1st Champion Value Prediction (CVP-1)*, held in conjunction with ISCA, June 2018. 98

[194] Y. Ishii, Context-base computational value prediction with value compression, in *1st Champion Value Prediction (CVP-1)*, held in conjunction with ISCA, June 2018. 98

[195] K. Koizumi, K. Hiraki, and M. Inaba, H3VP: History based highly reliable hybrid value predictor, in *1st Champion Value Prediction (CVP-1)*, held in conjunction with ISCA, June 2018. 98

[196] Y. Sazeides and J. E. Smith, The predictability of data values, in *International Symposium on Microarchitecture (MICRO)*, December 1997. DOI: 10.1109/micro.1997.645815. 98

[197] B. Reagen, J. M. Hernández-Lobato, R. Adolf, M. Gelbart, P. Wahtmoug, G.-Y. Wei, and D. Brooks, A case for efficient accelerator design space exploration via bayesian optimization, in *International Symposium on Low Power Electronics and Design (ISLPED)*, July 2017. DOI: 10.1109/islped.2017.8009208. 98

[198] W. Wang, J. W. Davidson, and M. L. Soffa, Predicting the memory bandwidth and optimal core allocations for multi-threaded applications on large-scale NUMA machines, in *International Symposium on High-Performance Computer Architecture (HPCA)*, March 2016. DOI: 10.1109/hpca.2016.7446083. 98

[199] A. Vallero, A. Savino, G. Politano, S. D. Carlo, A. Chatzidimitriou, S. Tselonis, M. Kaliorakis, D. Gizopoulos, M. Riera, R. Canal, A. Gonzalez, M. Kooli, A. Bosio, and G. D. Natale, Cross-layer system reliability assessment framework for hardware faults, in *International Test Conference (ITC)*, November 2016. DOI: 10.1109/test.2016.7805863. 99

Authors' Biographies

LIZHONG CHEN

Lizhong Chen is an Associate Professor in the School of Electrical Engineering and Computer Science at Oregon State University. He received his Ph.D. in Computer Engineering and M.S. in Electrical Engineering from the University of Southern California in 2014 and 2011, respectively, and B.S. in Electrical Engineering from Zhejiang University in 2009. His research interests are in the board area of computer architecture, interconnection networks, GPUs, machine learning, hardware accelerators, and emerging IoT technologies. Dr. Chen is the recipient of National Science Foundation (NSF) CAREER Award, several Best Paper Awards/Nominations at major architecture conferences, Chu Kochen Award (the highest honor from Zhejiang University), and an inductee of the HPCA Hall of Fame. He is also the founder and organizer of the Annual International Workshop on AI-assisted Design for Architecture (AIDArc), held in conjunction with ISCA. Dr. Chen is currently serving on the editorial board of *IEEE Transactions on Computers* (TC) and, in the past, has served as a program committee member in major computer architecture conferences (e.g., ISCA, HPCA, MICRO, DAC, ICS, IPDPS, IISWC), reviewer for several IEEE and ACM journals (e.g., TC, TPDS, TVLSI, TCAD, TACO), and panelist of multiple NSF panels related to computer systems architecture. He is a Senior Member of IEEE and ACM.

DREW PENNEY

Drew Penney is currently a Ph.D. student at Oregon State University and is a member of the System Technology and Architecture Research (STAR) Lab, directed by Dr. Lizhong Chen. At the STAR Lab, Drew explores novel machine learning applications to diverse architectural designs. He received his Bachelor's degree in Electrical and Computer Engineering (Summa Cum Laude) from Oregon State University and was a Dean's Scholar. He has published several papers on AI in computer architecture, was an invited guest speaker at a workshop on AI-assisted Design for Architecture (AIDArc), and received the Best Paper Runner-up Award in HPCA 2020 for his work on deep reinforcement learning in network-on-chip design.

DANIEL JIMÉNEZ

Daniel Jiménez is a Professor in the Department of Computer Science and Engineering at Texas A&M University. He was previously Assistant and later Associate Professor in the Department of Computer Science at Rutgers University, and Professor and Chair of the Department of Computer Science at UT San Antonio. Dr. Jiménez received his Ph.D. in Computer Sciences from UT Austin in 2002. He is interested in characterizing and exploiting the predictability of programs to improve microarchitecture. He pioneered the development of neural-inspired branch predictors, which have been implemented in microprocessors from AMD, Oracle, and Samsung. Dr. Jiménez designed the neural branch predictor for the Samsung Exynos M1 which is used in the popular Samsung Galaxy S7. He is a Senior Member of the IEEE, an ACM Distinguished Scientist, an NSF CAREER award winner, and member of the HPCA and MICRO halls of fame. He was the General Chair of IEEE HPCA in 2011, Program Chair for IEEE HPCA in 2017, and Chair of IEEE Technical Committee on Computer Architecture (TCCA) in 2018.

Printed in the United States
by Baker & Taylor Publisher Services